T0316322

The Regulatory Approach to Air Quality Management

In the wake of the Clean Air Amendments of 1970 in the United States, sources of emissions could be held accountable for the degradation of air quality in the local environment. This case study of air quality management in New Mexico was produced to shed some light on the procedures and activities used by agencies in order to control air quality. Originally published in 1981, Winston Harrington uses New Mexico as a case study for its largely centralised control system in Santa Fe to explore the behaviour of air quality agencies and pollution sources and comments on policy implications from this study's conclusions. This title will be of interest to students of environmental studies and policy makers.

The Regulatory Approach to Air Quality Management

A Case Study of New Mexico

Winston Harrington

RFF PRESS
RESOURCES FOR THE FUTURE

First published in 1981
by Resources for the Future, Inc.

This edition first published in 2016 by Routledge
2 Park Square, Milton Park, Abingdon, Oxon, OX14 4RN
and by Routledge
711 Third Avenue, New York, NY 10017

Routledge is an imprint of the Taylor & Francis Group, an informa business

Publisher's Note
The publisher has gone to great lengths to ensure the quality of this reprint but points out that some imperfections in the original copies may be apparent.

Disclaimer
The publisher has made every effort to trace copyright holders and welcomes correspondence from those they have been unable to contact.

A Library of Congress record exists under LC control number: 81081368

ISBN 13: 978-1-138-94455-8 (hbk)
ISBN 13: 978-1-315-67180-2 (ebk)

Research Paper R-25

The Regulatory Approach to
Air Quality Management
A Case Study of New Mexico

Winston Harrington

RESOURCES FOR THE FUTURE / WASHINGTON, D.C.

RESOURCES FOR THE FUTURE, INC.
1755 Massachusetts Avenue, N.W., Washington, D.C. 20036

Resources for the Future is a nonprofit organization for research and education in the development, conservation, and use of natural resources and the improvement of the quality of the environment. It was established in 1952 with the cooperation of the Ford Foundation. Grants for research are accepted from government and private sources only if they meet the conditions of a policy established by the Board of Directors of Resources for the Future. The policy states that RFF shall be solely responsible for the conduct of the research and free to make the research results available to the public. Part of the work of Resources for the Future is carried out by its resident staff; part is supported by grants to universities and other nonprofit organizations. Unless otherwise stated, interpretations and conclusions in RFF publications are those of the authors; the organization takes responsibility for the selection of significant subjects for study, the competence of the researchers, and their freedom of inquiry.

Research Papers are studies and conference reports published by Resources for the Future from the authors' typescripts. The accuracy of the material is the responsibility of the authors and the material is not given the usual editorial review by RFF. The Research Paper series is intended to provide inexpensive and prompt distribution of research that is likely to have a shorter shelf life or to reach a smaller audience than RFF books.

Library of Congress Catalog Card Number 81-81368

ISBN 0-8018-2700-0

Copyright © 1981 by Resources for the Future, Inc.

Distributed by The Johns Hopkins University Press,
   Baltimore, Maryland 21218

Manufactured in the United States of America

Published May 1981. $7.50

TABLE OF CONTENTS

LIST OF TABLES

LIST OF FIGURES

PREFACE

The study reported in this volume was done as part of a program of research on implementation incentives for environmental quality management under the auspices of the U.S. Environmental Protection Agency, Charles N. Ehler, Project Officer, with the cooperation of the U.S. Council on Environmental Quality, Edwin H. Clark, Project Officer.

New Mexico was chosen for this case study in part because it offered the virtue of simplicity. There are only about one hundred stationary sources in the state; in contrast, neighboring Texas has over a thousand. Likewise, the administrative structure of the air quality division of the Environmental Improvement Agency (EIA) is straightforward and largely centralized in Santa Fe. This made it possible, in a relative short time, to gain a reasonable understanding of the problems and procedures of air quality management in New Mexico. Notwithstanding the apparent simplicity of the situation, the problems of enforcement and the attempted solutions seem representative of the process and problems in larger and more complex states. It should be emphasized that the objective was in no way to evaluate the state program; I was interested in _how_ it works rather than _how well_.

The EIA keeps reasonably complete files containing all the correspondence between it and each major stationary source in the state. These files, together with the records of permits and variances granted and interviews with EIA and source personnel, constitute the data base for this analysis.

In the course of the study I spent approximately three weeks in Santa Fe at the Environmental Improvement Division of the State Department of Health and Social Services. (A government reorganization during the spring of 1978 made the free-standing Environmental Improvement Agency [EIA] a unit of this omnibus state health department. In this report the state air quality agency is referred to as the EIA, because that title was used during the period of interest, 1970-1978).

The work could never have begun without the cooperation and assistance of the Air Quality Division of the New Mexico Environmental Improvement Agency. Its director, Ken Hargis, graciously allowed me to spend three weeks in his department, even providing me with a private office to use as a base. I owe a particular debt to David Duran, the head of the Enforcement Section, who suffered with the utmost patience and good humor my interminable questions, and Gary Kittams, who spent a lot of time with me on the phone after I returned to Washington.

I received many helpful comments on an earlier draft from David Abbey of Los Alamos Scientific Laboratory and Bruce Aronson, Blair Bower and David Goetze of RFF. Mr. Bower in particular offered numerous helpful comments on content and organization. Excellent editorial advice was received from Kent Price and Sally Skillings. I am also grateful for support from RFF for revision of the report for publication. Finally, I would like to thank Eileen LesCallett, Lee Carlson and Mae Barnes for cheerfully typing the manuscript and its numerous revisions.

Winston Harrington

Chapter 1

INTRODUCTION

The Clean Air Amendments of 1970 were a watershed in Federal efforts to improve air quality. For the first time, mechanisms were specified whereby sources of emissions could be held accountable for their contribution to the degradation of air quality. Federal and state agencies were delegated responsibility to devise and enforce stiff emission standards which would result in the achievement of ambitious air quality goals.

As a result, it is estimated that $279 billion will be spent for controlling air pollution between 1978 and 1987, split about 40-60 between mobile and stationary sources.[1] This vast commitment of resources naturally makes the Clean Air Act a very controversial piece of legislation. But so far the controversy over clean air policy has focused on the questions of whether the reduced emissions will result in improvements in air quality, and whether the benefits of improved air quality are worth the cost. Usually it is assumed that these expenditures will result in the anticipated reduction in pollution discharge.

The validity of this assumption depends, at least in part, on the effective enforcement of these emission standards by local and state air

----

[1] U.S. Council on Environmental Quality, Seventh Annual Report, Washington, D.C., 1976, p. 666.

quality agencies and by the U.S. Environmental Protection Agency (EPA). If this assumption is not valid, that is, if despite the best efforts of EPA and the state agencies substantial noncompliance with air quality regulations occurs, the benefits expected from this rather large commitment of resources to air quality may not fully materialize.

Very little research has been done on the enforcement activities of air quality agencies and the response of sources to those activities. To help shed light on these issues a case study of air quality enforcement in New Mexico was undertaken. This paper reports on the findings of that study.

The Context of Air Quality Management

Governments at three levels--federal, state, and local--have some responsibility for achieving and maintaining ambient air quality in New Mexico, or indeed in any state, although it was not always this way. Traditionally, what little concern there was with air quality was found at the local level only, where air pollution was dealt with both by ordinance and the common law. It would not be correct to say that locally-inspired air pollution policy was completely ineffectual. In New Mexico, several asphalt plants as well as other types of sources which happened to be located in populated areas installed flue gas cleaning devices long ago. Presumably they did this to avoid a nuisance action or simply to maintain good community relations, but in either case, a positive action was taken to preserve air quality out of a concern over possibly unfavorable local reaction. Nonetheless it was perceived that exclusive reliance on local government was not preventing air quality problems from getting worse, and, in the mounting concern over environmental quality, Congress responded with the Clean Air Act of 1963 (P.L. 88-206), later amended by the Air Quality Act of 1967 (P.L. 90-148) and the Clean Air Amendments of 1970 (P.L. 91-604).

The 1963 Clean Air Act was principally concerned with developing scientific data on the effects of air pollution, although federal action was allowed for where public health or welfare was jeopardized. The 1967 act significantly strengthened the federal role by requiring the states to establish ambient air standards and implementation plans for meeting them. But this Act contained no effective federal enforcement mechanism, and the 1970 Clean Air Act was intended above all to remedy this problem. In effect, the Act called for a national plan to achieve and maintain throughout the U.S. ambient air quality standards (NAAQS)[2] sufficient to protect the public health (the primary standards) and welfare (the secondary standards) (Sec. 109).

In 1971 the EPA established NAAQS for six pollutants. The states were then required to prepare state implementation plans (SIPs), specifying emission limitations and a program for their enforcement (Sec. 110). The implementation of the SIP would lead to the achievement of national ambient air quality standards in that state. All the SIPs had to be approved by EPA, and in the event that a state could not or would not prepare an acceptable plan, EPA was to prepare and enforce an implementation plan for the state. For a plan to be acceptable, EPA had to be satisfied that the state's implementation would in fact lead to achievement of the AAQ standards within three years for the primary and a "reasonable time" for the secondary standards. In

---

[2] States were allowed to set their own ambient standards, as long as they were at least as stringent as the federal standards.

addition, the state had to demonstrate that it had both the resources and the legal authority necessary to carry out the plan. This meant, among other things, that the state had to show that it had the ability to set and enforce emission limitations for stationary sources.

The actual implementation of an EPA-approved plan was a joint responsibility of all levels of government, though it was supposed to be the "primary responsibility" of the states (Sec. 107(a)). Local governments could get involved in two ways. First, as noted above, many localities had existing air pollution regulatory programs, and these were usually expanded to help implement state plans in their regions. In addition, the Amendments called for greater use of "Air Quality Control Regions," first defined in the 1967 Act, for purposes of developing and carrying out implementation plans (Sec. 107(a)). In the event that the plan or its implementation appeared inadequate, EPA could assume responsibility (Sec. 113(a)(2)). Furthermore, EPA was involved not only through its approval of state plans but also through direct regulations of sources.[3] Any provision of a SIP that was approved by EPA became a federal regulation, so that even if a state were unwilling or unable to act against a violator of a regulation, after a 30-day notice EPA could intervene and enforce it (Sec. 113(a)(1)).

Within this context the successful implementation of air quality programs depends on the interactions between air quality agencies and the sources they regulate, as well as the interaction among air quality agencies

---

[3] The act also directed EPA to establish and enforce other source-specific regulations, such as the New Source Performance Standards (Sec. 111).

at the various governmental levels. This case study concentrates on inter-

actions in the first category--specifically, the relationship between the

Air Quality Division of the New Mexico Environmental Improvement Agency

(EIA) and the air pollution sources in that state.

Two related concerns dominate the discussion. First, what procedures

are used to enforce the regulations, and what activities are performed by

agency personnel? Successful enforcement of air quality regulations

requires that the agency perform four functions: It must have a definition

of compliance, it must be able to detect violations, it must be able to

bring noncomplying sources back into compliance, and it is certainly desir-

able if the agency has means of discouraging noncompliance from occurring

in the first place. To what extent do agency procedures and activities

promote or hinder performance with respect to these tasks? What costs are

imposed by enforcement activities? Included here are the administrative

costs borne by both public agencies and regulated sources and the cost to

the sources of complying with the regulation.

The second concern is with the response of sources to the regulations

and their enforcement. What choices are available to the sources? On what

basis does a source choose its response?

As noted above, this report does not deal with the issue of federalism.

However, the interaction between EPA and the state agency conditions--and

is conditioned by--the relations between the state agency and the sources.

Accordingly, it is useful to discuss briefly the EPA-state relationship.

The essence of that relationship is that a national objective--achievement

of the national ambient air quality standards--is to be implemented largely

through actions taken by local and state governments. Presumably this means that EPA must have ways of inducing the states to take actions they would not take in the absence of federal legislation; otherwise there would have been no need for a Clean Air Act. Therefore, just as the state agencies must provide implementation incentives to air pollution sources, EPA must provide incentives to the state agencies.

One such incentive is that any part of a SIP approved by EPA becomes a federal regulation. The main purpose of this provision is to ensure that air quality management does not fail simply because of state unwillingness or inability to operate an effective program. However, it could also be expected to provide an incentive to the state government to maintain an effective program. Most states would rather administer their own programs rather than have EPA managing their air resources. In addition, the fact that EPA stands behind the state ready to intercede may help the state agency resolve political problems over regulations and their enforcement. The agency can point to the presence of EPA looming in the background and argue that if the agency does not take a hard line then EPA will. Of course, intercession by EPA does not have to apply to the program at large. It can be and usually is directed at violations of a particular regulation or even the emissions of a particular source. Obviously, if it could only be directed at an entire program it would not be nearly as effective an incentive.

The other incentive that EPA can provide to state and local agencies is in the matter of financial aid. EPA provides substantial grants to state and local air pollution control agencies for developing and maintaining air pollution programs (Sec. 105). In New Mexico, EPA currently provides about one-half of the budget of the Air Quality Division of the EIA, and similar support can be expected in future years. These grants act as an incentive to the state in at least two ways: (i) they make it cheaper for the state to allocate resources for the state program, especially if they are offered according to some matching formula, and (ii) there is always an implied threat of losing the support if the operation of the program is deemed unsatisfactory.

Some Definitions

In this section definitions of a number of terms used frequently in this report are given.

A source is an individual discharge location within a plant, such as a stack, pump or storage tank. These discharge points are often designed specifically to discharge waste products into the atmosphere. Emissions from sources not in this category are called fugitive emissions. Examples of sources of fugitive emissions include storage piles, dirt roads, and leaky equipment. Clearly, a plant may contain several sources, and likewise a firm may encompass several plants.

Shortly after source-specific regulations were established in the early 1970's, sources had to demonstrate their ability to meet them.[4] Similarly, new plants were required to pass a performance test shortly after beginning operation. Sources successful in this regard were then in initial compliance with the regulations. However, it was entirely possible for a source to be in initial compliance and yet subsequently exceed the applicable emission standards because of a failure to operate the emission

---

[4]Under some circumstances sources were given variances which enabled them to delay the date of compliance with the regulation. See Chap. 4.

control system properly. Thus it was important to keep sources in <u>con-</u>
<u>tinuous</u> compliance with the regulations as well.

Air quality agencies gain information about continuous compliance from
both source <u>surveillance</u> and from <u>self-reporting</u>. Types of surveillance
activities conducted by the agency include a <u>source test</u>, in which an estimate of pollutant discharge is made from a sample of stack emissions,
<u>opacity readings</u>, in which a trained observer estimates the optical density
of the stack plume, and an <u>engineering inspection</u>, in which an agency
engineer makes an assessment of a plant's compliance status without sampling
stack emissions. Often sources are also requested or required to provide
their own data on emissions. For example, sources are occasionally asked
to conduct source tests and report on the results. In addition, some types
of sources are required to submit periodic reports on emissions calculated
from a <u>materials balance</u>, in which emissions are estimated as the amount of
pollutant in raw material inputs less the amount in product and by-product
outputs.·

Organization of the Report

This report can be divided into three parts: institutional context
(Chapters 2 and 3), agency activities and source response (Chapters 4, 5
and 6), and conclusions (Chapter 7). Chapter 2 contains a discussion of
the organizations responsible for air quality enforcement in New Mexico,
their relationship to each other, and their relationship to the federal
EPA. Some background material on New Mexico's air quality regulations is
provided in Chapter 3. The chapter begins with a brief examination of the
new source permitting process. Then the emission regulations themselves
are described for each category of stationary sources in the state.

The heart of the study is contained in Chapters 4 through 6, which respectively examine the options available to a source seeking relief from a regulation, the enforcement practices of the state agency, and the compliance of sources with the regulations. Finally, some tentative conclusions are offered in Chapter 7. After a short discussion of the generalizability of the results, some hypotheses about the behavior of both air quality agencies and pollution sources in this sort of regulatory setup are stated. This is followed by an exploration of the policy implications of those hypotheses.

Chapter 2

INSTITUTIONAL BACKGROUND

New Mexico has two state government organizations with environmental responsibility: the Environmental Improvement Board (EIB) and the Environmental Improvement Division of the Department of Health and Environment. In addition, there is a set of local environmental agencies for Bernalillo County (containing Albuquerque). In this chapter the functions of and relationships among these agencies are described.

State Air Quality Management

The Environmental Improvement Board was created by the Environmental Improvement Act of 1966 (NMSA Sec. 12-12-1). Its principal responsibility is to promulgate regulations in a variety of areas related to public health and welfare, including air and water quality management, solid waste, water supply, food protection, consumer product safety, radiation control and occupational health and safety. It has been designated the state air pollution control agency for all purposes of the federal Clean Air Act.

The EIB consists of five members appointed by the governor, subject to approval by the state senate, in overlapping terms of five years each. The statute does not specify any particular qualifications required of board members, except that no more than three can be of the same

political party. Three members constitute a quorum, although any action
by the board requires the concurrence of three members present at a
meeting.

With regard to air quality management the EIB's duties include the
following: (i) approval of the State Implementation Plan (SIP), (ii) the
determination of ambient air standards, (iii) the adoption of emission
standards for stationary sources, (iv) the granting of variances and the
acceptance of "assurances of discontinuance" (both are legal devices which
allow a source to discharge emissions in excess of applicable regulations
under certain conditions), and (v) the hearing of appeals of permit
denials. Duties (iv) and (v) above make it clear that the EIB has judicial as well as legislative responsibilities.

The Environmental Improvement Division was known as the Environmental
Improvement Agency (EIA) until 1978 when a reorganization of
state government placed it in a cabinet-level department. Because throughout the period we are concerned with it was known as the Environmental
Improvement Agency, I refer to it by that title in this report. Like
the EIB, the EIA was created by the Environmental Improvement Act in 1966.
Its main function is the administration of the regulations promulgated by
the EIB. The air quality responsibilities of the EIA include the following: (i) preparation of the SIP, (ii) enforcement of emission standards,
(iii) issuance or denial of permits, and (iv) acting as staff to the Board.

The importance of the Environmental Improvement Agency is actually
greater than this discussion might indicate, because of its role as
advisor to the Board. Before taking any action the Board is furnished
with a recommendation from the EIA. When exercising its legislative

functions, the board ordinarily acts on proposals presented by the EIA

(competing or alternative proposed regulations may also be submitted by

other interested parties). In its judicial capacity (deciding on vari-

ances or assurances of discontinuance) the board is furnished with an

agency recommendation before taking any action. It is quite rare for the

board to go against the agency recommendation, although not unheard of.

Thus, the reason for the agency's influence is that it has all the informa-

tion. The board has no independent staff, and must rely on the agency for

staff work. Inasmuch as it controls the flow of information, the agency's

recommendations carry much weight.

The Air Quality Division in the Environmental Improvement Agency has

a staff of 35, divided into four sections. The SIP Section is responsible

for keeping up to date the State Implementation Plan and is also respon-

sible for disposing of permit applications. The Meteorology Section has

the responsibility for modeling ambient air quality. The

Ambient Air Monitoring Section must maintain the ambient monitors

in the state and report data to EPA. Finally, the Enforcement Section is

composed of 7 professionals, all technical, whose primary activity is

source surveillance. The Enforcement Section also has the principal

responsibility for making recommendations to the board about variances

and assurances of discontinuance.

Air Quality Management in Bernalillo County

Air Quality Management in New Mexico began in Bernalillo County when,

in 1965, Albuquerque established its own air quality program. Two years

later the State Air Quality Control Act was passed, using the Albuquerque

program as a model. The city government has thus been a leader in the
state where air quality was concerned.

In the early seventies, however, the air quality program in Albuquerque fell upon hard times due to a lack of political support. The city
administration at the time took the position that air quality management
was a state government function. One of the provisions of the State Air
Quality Control Act states that if the local agency fails to administer
its program, the state EIA may take it over, and the city was attempting
to force the state to do just that through its neglect. But the state at
that time had its hands full in trying to visit every other stationary
source in the state, and was reluctant to take on the additional responsibility.

During this period there was very little source surveillance or
other enforcement activity for stationary sources. Source surveillance
was the responsibility of one person, and most of his time was spent in
"walk-throughs" of plants. A walk-through is one cut below an engineering
inspection, in which the inspector looks for casual evidence of compliance
or noncompliance without, for example, measuring the flow rate or pressure
drop in a scrubber. With respect to source tests, the local agency did not
have the facilities to perform them according to EPA Method 5.[1] Instead

---

[1]To ensure consistency and, presumably, accuracy EPA has specified
a number of stack test methods, which are to be used by sources to demonstrate compliance with the federal New Source Performance Standards, in
the appendices to 42 CFR Chapter 60. These methods have been adopted in
New Mexico.

it used equipment designed for quick estimates of emission rates, but which sacrificed some accuracy compared to EPA Method 5. In any case, source tests were rare events, having occurred at most four times prior to 1977. In fairness, it must be pointed out that air quality management was not entirely moribund in Bernalillo County at this time. The local agency did manage against heavy opposition to extend regulation to hydrocarbon emissions from tank farms.

Nonetheless, EPA Regional Headquarters in Dallas was growing increasingly dissatisfied with the performance of the local agency and doubted the commitment of the city to air quality management. The Region expressed this dissatisfaction by refusing to increase the federal grant contribution to the local program. For five years between 1973 and 1977 the grant remained constant, during which time the EPA share in the local budget declined from 67 to 33 percent. Finally, in late 1977, EPA threatened to withdraw support entirely unless the program was improved, especially with respect to enforcement.

By this time Albuquerque had a new city administration, and the new mayor promised in early 1978 that the deficiencies in the program would be addressed. The administration brought in the assistant chief of the Air Quality Section of the EIA to head the rejuvenated program. At this time, all the stationary sources in Bernalillo County were listed on the CDS system as being in compliance.[2] Under the new regime, however, all sources

---

[2]CDS is EPA's Compliance Data System, a computer program which reports on the compliance status of every major source in the U.S. States are required to report on the compliance status of all sources twice a year, and these data are input to CDS.

were designated in April as having unknown compliance status. The local

agency requested help from the state EIA in ascertaining the compliance

status of the stationary sources in Albuquerque, and in June an inspection

team from Santa Fe performed engineering inspections on seven sources in

Bernalillo County (4 construction companies, 2 tank farms and a perlite

plant). Two of the construction companies were found to be out of compli-

ance. Due to manpower limitations the EIA will only be able to offer

occasional assistance to Bernalillo County in determining compliance. Some

of the testing will probably be done by the staff of the local agency, while

for the rest the agency will probably request source tests or other proof of

compliance from the sources themselves.

The organization for air quality management in Bernalillo County is

similar to that found at the state level. The Albuquerque-Bernalillo County

Air Quality Control Board (AQCB) is a judicial/legislative body analogous to

the State EIB. Enforcement power is vested in the AQCB but delegated to the

City-County Department of Environmental Health, which corresponds to the

EIA. But though organization is similar there are important differences

with respect to the problem faced. The main difference is that in Albuquer-

que one finds much greater concern with mobile than stationary sources.

Like a number of other cities in the Southwest, Albuquerque combines the

attributes of low density, so that there is a high reliance on automobile

transportation, and high altitude, so that one expects relatively higher

emissions per vehicle-mile. On the other hand, because there is little

heavy industry in the city, stationary sources are not so important. Among

the seventeen major stationary sources, one finds nine asphalt plants and

four nonmetallic mineral plants. In fact, much of the work done with respect to stationary sources is not directed at these major stationary sources at all. The city is equally interested in hydrocarbon emissions from gas stations when their underground tanks are being replenished, and the regulations require the recycling of vapor.

Administratively, the practical relationship between the EIA and the Department of Environmental Health is one of equals rather than that of superior and subordinate, although as a legal matter the state is ultimately responsible. The local agency deals directly with the regional EPA. About the only connection is that the EIA occasionally provides the local agency with technical assistance. For this reason, and also because of the low level of enforcement activity by the local agency for most of the past decade, I have concentrated in this report on enforcement activities by the EIA only.

Chapter 3

REGULATORY BACKGROUND

This chapter discusses the structure of regulations applicable to stationary sources in New Mexico. The first section deals with the registration of sources and the permit process for new sources. The second describes briefly the categories of stationary sources found in New Mexico and the regulations specific to those sources. The third section describes a set of other regulations applicable to all sources.

Permits

New Mexico requires the owner or operator of any commercial or industrial stationary source with potential (uncontrolled) annual emissions of one ton or more of any air contaminant to register with the EIA.[1] In effect this requirement includes every stationary source in the state.

To get a registration certificate, a source must furnish the EIA with information describing the nature and amounts of pollution generated, the processes and air pollution control equipment used, and expected operating schedule. The registration requirement is for informational purposes

---

[1]"Air contaminant" is defined by statute (N.M.S.A. Sec. 12-14-2) to be "any substance, including but not limited to any particulate matter, fly ash, dust, fumes, gas, mist, smoke, vapor, micro-organisms, radioactive material, any combination thereof or any decay or reaction product thereof."

only; the department cannot refuse to register a source unless the application is incomplete. Principally it is a device through which the EIA could get emission information from existing sources when the state was first setting up its regulatory program.

For a new source—or modification of an existing source—with potential emissions in excess of either 10 lbs per hour or 25 tons per year,[2] or any emissions of a hazardous pollutant, a permit is required.[3]

This permit is a construction permit only and must be obtained before the source can begin construction. Unlike many other states, New Mexico does not require a renewable operating permit of any source. Also unlike some other states, no fee is required to cover the cost of processing the permit. The amount of Agency effort required to process an application varies tremendously depending on the complexity of the permit, but the average is about two man-weeks—about $1,000.

The permit application procedure is specified in Air Quality Regulation 702. As shown in Figure 3-1, the source first applies for a permit or registration certificate (both are made on identical forms). If potential annual emissions are 25 tons or less, or if the maximum emission rate is 10 pounds per hour or less, the EIA immediately issues a registration certificate.

---

[2] Thus permits are required of considerably more than just "major sources," a term used by EPA for any source with emissions in excess of 100 tons per year.

[3] Hazardous air pollutants are those designated as hazardous air pollutants by EPA pursuant to the Clean Air Act of 1970.

Figure 3-1 New Mexico's Air Quality Permit Process

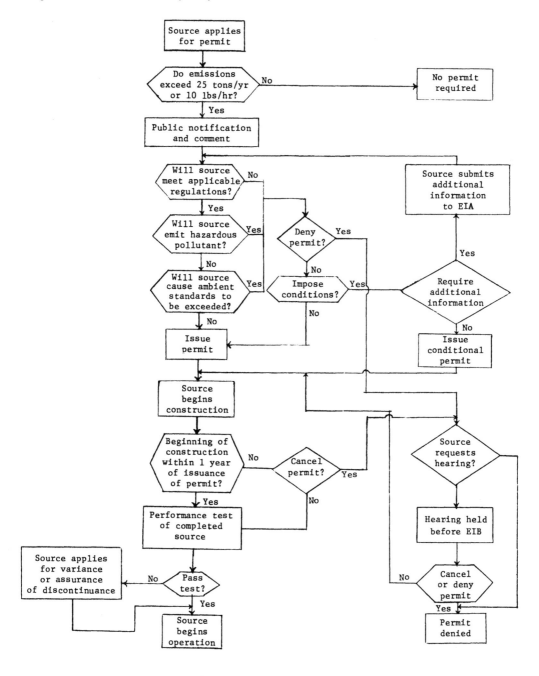

Otherwise the department begins the processing of the permit application. It first makes a public announcement of the receipt of the application by a news release to the major state papers for large sources and to local papers for smaller sources. The announcement contains a preliminary determination by the agency regarding whether a permit will be issued, and invites comment by interested parties to be submitted within 15 days of receipt of application.

In evaluating the permit application the EIA considers above all (i) whether the source will meet applicable regulations, (ii) whether the source will emit a hazardous pollutant, and (iii) whether the source will cause a violation of ambient standards, using the information contained in the permit application and that submitted during the comment period. As shown in Figure 3-1, the EIA may choose to gather additional information from the source, to deny the permit, or to issue a conditional or unconditional permit. A "yes" answer to one of the questions above is usually enough to cause conditions to be placed on the permit, but it is not a statutory requirement. Issuance of an unconditional permit, of course, does not exempt the source from compliance with the applicable regulation. One of the conditions the EIA may require in a conditional permit is the submission of additional evidence of the environmental impact of the source. After the permit has been issued, the source has one year to begin construction, or the agency at its discretion may cancel it. A source may appeal any permit denial, cancellation, or condition to the Environmental Improvement Board.

The final phase of the permit application procedure is the performance test of the completed source. This test may be performed either by an Agency source testing team or by a contractor retained by the source. If the test is failed, then the source must seek a variance or assurance of discontinuance to continue to operate while it attempts to find the trouble.

In New Mexico the permit requirement is not so much an enforcement tool as an information-gathering tool. It tells the department who and where the sources are and what their emissions are likely to be, at least under ideal conditions. This is essential information for ambient air quality modeling. It is very rare for permit applications to be denied.

In order to get a better idea of permit procedures, all permit applications received between July 1976 and August 1978 were examined. During this period 102 permit applications were received, about one per week, and only one was denied. Fifteen other applications were held up because the department needed additional information, and for another ten it was determined that no permit was required.

By regulation the department is supposed to make a disposition on permits within 30 days of receiving them; however, the department has had trouble meeting this goal. The median processing times of three categories of applications are given in Table 3-1.

As shown, the median time required to process "routine" applications (i.e., those not returned for additional information) was 37.5 days, and 25 percent took two months or longer. The principal reason for this

Table 3-1. Processing Times for Permit Applications, 9/76-8/78
            (Calendar Days after Receipt of Application)

	Median	Interquartile Range	Number of Applications
Applications not requiring permit	13.5	5.5 - 18	10
Applications returned for more information	57	43 - 73	16
All other applications	37.5	29.5 - 59	76

excess--one not usually contemplated when regulations of this sort are

written--is the fact that permit arrivals do not arrive at a constant rate.

There is, first of all, an apparent seasonal pattern; over the sample

period the number of permit applications received was as follows:

Quarter	Applications	Quarter	Applications
III-76	7	III-77	9
IV-76	14	IV-77	16
I-77	12	I-78	11
II-77	8	II-78	7

Presumably, the permit application rate is high in the winter months

because sources want to make sure they can begin construction the following

spring. Even within a quarter, permits arrive erratically. In the winter

of 1977-78, for example, 11 permits were received in December, two in

January, and nine in February. The EIA can increase its processing rate

when necessary, but it cannot match the variation in permits received.

The average backlog in permit applications is five but it has gotten

as high as ten. Given an average processing time of one week, the average

elapsed time for processing an application is five weeks. It cannot

be shortened without devoting additional resources to permit applications.

But even though the time required slightly exceeds the statutory limit,

permit applicants have not complained.

Major Stationary Sources in New Mexico and the Regulations Pertaining to Them

New Mexico contains about 120 "major stationary sources" (those with

potential emissions in excess of 100 tons per year). The exact number

fluctuates, principally because of portable asphalt processing plants,

which are set up and moved around the state to work on specific paving

jobs. These major sources are found in ten different industries as shown

in Table 3.2. The sources in Bernalillo County have been separated out

because the local health department has enforcement authority there, while

the state Environmental Improvement Agency has responsibility elsewhere in

the state.

In this section we provide some background information for each of

these categories with respect to the pollutants generated and the regula-

tions to which they are subject.

For Natural Gas Processing Plants. Natural gas in New Mexico is pro-

duced in the Permian Basin in the southeast corner of the state, and in the

San Juan Basin in the northwest. Permian Basin gas is "sour gas," contain-

ing appreciable amounts of hydrogen sulfide which must be stripped before

Table 3-2. Industrial Stationary Sources in New Mexico as of April 1, 1978.

Industry	Number of Plants			Pollutants Regulated
	Bernalillo County	Rest of State	Total	
Natural gas processing	0	38	38[a]	SO_2, Total sulfur emissions
Petroleum refining	0	7	7	SO_2, HC, 7 other pollutants
Utility steam generators				
Coal-fired	0	3	3	SO_2, NO_x, particulates
Gas-fired	2	8	10	NO_x
Oil-fired	0	2	2	SO_2, NO_x, particulates
Asphalt processing	9	27	36[b]	Particulates
Woodwaste burners	0	10	10	Smoke (opacity)
Sulfuric acid plants	0	2	2	SO_2
Nonferrous smelters	0	2	2	SO_2, particulates
Non metallic minerals				
Mica, perlite, pumice	0	7	7	Particulates
Gypsum	1	1	1	Particulates
Cement	1	0	1	Particulates
Gravel	2	0	2	Particulates
Total Major Sources	15	105	122	

[a] Only 8 plants are affected by sulfur emission regulations.

[b] This is the number of firms; several firms have more than one processing plant.

the gas is allowed to enter the pipeline. New Mexico regulations require that this gas be oxidized to SO_2 before it is vented to the atmosphere. Furthermore, those plants which liberate 7.5 tons per day or more of sulfur are required to recover at least 88 (or in some cases 90) percent of input sulfur. Eight processing plants in New Mexico have sulfur recovery systems. Most had to install these units in order to come into compliance with the regulation, though at least one gas plant, a sulfur recovery unit, was installed in the sixties because it was profitable to do so at the time. However, the conditions which gave rise to this unit have disappeared with the rise in the cost of construction of such units and the decline in the real price of sulfur.

For Petroleum Refineries. The state contains seven small refineries with a total of 118,000 barrels per day crude thruput. Pollutants generated at these refineries that are subject to emission standards include total sulfur, hydrocarbons, carbon monoxide, particulates and mercaptans. In practice, however, virtually all the regulatory activity has involved sulfur and hydrocarbon emissions. The principal impact of the regulations was to force the installation of sulfur and hydrocarbon recovery systems. Once installed, the operating expenses of vapor recovery are exceeded by the value of the recovered hydrocarbon, so that the major regulatory effort is to see to it that the systems are installed. Very little subsequent surveillance is required.

Such is not the case with sulfur emissions. State regulations require that sulfur emissions from existing refineries not exceed 5 tons per day. New plants processing crude containing up to 30 tpd sulfur are required to capture 90 percent of input sulfur while larger new plants are required to capture 98 percent. The regulations require that each plant report sulfur emissions on a monthly or quarterly basis.

For Asphalt Processors. Particulate matter is the only pollutant regulated for asphalt processors. The New Mexico regulation limits emissions to 10 to 50 lbs/hr, depending on output. In addition, no fugitive emissions are allowed; all emissions must exit through the stack. Most asphalt plants in the state are portable: they are set up for particular paving jobs, and dismantled when the jobs are concluded. An uncontrolled asphalt plant cannot meet state emissions regulations. Since the regulation became effective most plants have responded by installing scrubbers, although two plants in permanent locations installed baghouse filters. Plants used in urban areas typically had scrubbers installed before the regulations were established.

For Utility Steam Generators. There are two large coal-fired power plants in New Mexico, for which the state regulates emissions of particulates, SO_2, and NO_x. The emission standards are shown in Table 3-3. In addition there are thirteen smaller plants, ten of which were fired primarily with natural gas (959 MW), two primarily with oil (417 MW), and one small coal-fired plant (5 MW). The gas plants are subject to a NO_x emission regulation of 0.3 lbs per 10^6 Btu input. The oil plants must meet regulations on SO_2, NO_x and particulates of 0.34, 0.3, and 0.005 lbs per 10^6 Btu input respectively.

The standards for San Juan # 1 are much more stringent than the standards for San Juan # 2 or the Four Corners Units because it is classified as a new source. At San Juan # 2 equipment was recently installed which

Table 3-3. Emission Standards at the Four Corners and San Juan Power Plants

Plant	Capacity (MW)	Emissions (lbs/10^6 Btu Input)		
		Particulates	SO_2	NO_x
Four Corners (Arizona Public Service)				
Unit 1	175	0.135	1.0	0.7
Unit 2	175	0.135	1.0	0.7
Unit 3	225	0.135	1.0	0.7
Unit 4	800	0.5	1.0	0.7
Unit 5	800	0.5	1.0	0.7
San Juan (Public Service of New Mexico)				
Unit 1	330	0.05	0.34	0.45
Unit 2	330	0.135	1.0	0.7

enables the plant to meet the tighter standards on SO_2 and particulates applicable to Unit 1 The two units, in fact, are equipped with a very efficient hot-side precipitator and a Wellman-Lord regenerable scrubber, the latter being only the second such unit installed in the United States.

The Four Corners Plant, on the other hand, has with minor exceptions had to install no equipment to meet the state's regulations. In fact, the standards now in force were set more or less to be what the plant could achieve, without installing additional equipment, although within the next five years the plant will probably have to meet much more stringent standards. Note that the newer, larger units (4 and 5) at the Four Corners

Plant actually have less stringent particulate limitations than units
1-3. The regulations for units 4 and 5 will be discussed further in
Chapter 4.

For Woodwaste Burners. Twelve sawmills are found scattered around the
western half of New Mexico. Each creates a substantial quantity of woodwaste, mainly sawdust, chips, and bark. The traditional solution to the
woodwaste disposal problem is to burn it, usually in a "teepee burner,"
so called because of its conical shape. Potentially, this product can be
put to use as a mulch or in the manufacture of particleboard and one of
the responses to regulation has been a large increase in the utilization
of woodwaste for such purposes. Whereas prior to 1970 every saw mill in
the state burned all its woodwaste, today only three do. Of the remainder,
three have achieved full utilization[1] and seven are able to use part of the
waste. Most of the sawmills in New Mexico, however, are found in remote
locations, a fact which makes the economics of byproduct use marginal at
best.

The operation of a teepee burner produces smoke, the quantity of
which depends on the temperature at which the burner is operating. A
burner operating in such a way that the exit gas temperature exceeds 750°
F. produces very little smoke; emissions increase as the stack temperature
falls short of this figure.

---

[1] One of the three. however, utilizes its waste by burning to produce
steam, and this mill has been having trouble meeting the opacity regulation. It is owned by the Navajo Tribe and is located on the reservation.
The Tribe denies the state's jurisdiction over the mill, and the state is
refraining from prosecuting until the question of who has responsibility
for air quality management is settled in another court suit now pending.

Because there is no stack in which to measure emissions, the New
Mexico regulation for woodwaste burners is specified in terms of opacity.
All woodwaste burners in the state are required to maintain opacity at 20
percent or less, except for one hour during startup or burndown, when 40
percent opacity is permitted. Burners operating at night are required to
maintain temperatures in excess of 750° F., and to install temperature
recording devices for the purposes of demonstrating compliance with this
provision.

For Sulfuric Acid Plants. At New Mexico's two sulfuric acid plants
the principal regulation limits sulfur emissions to 5 to 7 percent of the
sulfur processed in the plant. In addition, any hydrogen sulfide released
must be flared to SO_2. These plants are both important potential sources
of SO_2, with over 100 tons per day of SO_2 produced (prior to treatment).
Both, however, seem to be able to meet the regulation easily.

For Nonferrous Smelters. New Mexico contains two copper smelters--a
Kennecott facility at Hurley and a Phelps-Dodge plant near Grants. The
air pollutants regulated are SO_2 and particulates. The Phelps-Dodge plant
is a new plant, having commenced operation in 1976. The regulations for
new plants require the capture of 90 percent of the sulfur released in the
plant, and for particulates the plant is limited to 0.03 grains per cubic
foot of stack gas for each stack at the plant.

The standards for Kennecott are much less stringent. For SO_2, only
60 percent control is required, although the regulation did force Kennecott
to install a sulfur recovery system. For particulates, the original regulation was amended in 1976 to approximate the plant's actual emissions,

although a hearing was held in 1978 by the EIB to discuss more stringent regulations.

For Nonmetallic Mineral Processors. There are a dozen plants in this category in the state. The mica, perlite, and pumice processors are all governed by the same regulation, which allows emissions on a sliding scale, from 10 lbs per hour for a processing rate of 5 tons per hour, up to 50 lbs per hour for a processing rate of 300 tons per hour. For gypsum, concentrations up to 690 mg/m^3 (0.3 grains/cubic foot) are allowed. This regulation is comparable to the perlite regulation; at one gypsum plant it translates into an emission rate of about 25 lbs per hour. Every one of the nonmetallic mineral processors in New Mexico uses a baghouse filter to control emissions. At at least one plant an electrostatic precipitator was tried, but it did not work very well and has been replaced by a baghouse.

Other Regulations. In addition to these source-specific regulations, there are other regulations of interest. The first is a regulation forbidding open burning except for certain purposes (including backyard barbecues, natural gas flaring, agricultural management, and refuse disposal in small communities). For certain other purposes, such as forestry management, open burning is allowed when a permit is obtained.

New Mexico also has a regulation to control smoke. This regulation requires that any source be required to limit its stack opacity to 20 percent (Ringelmann 1) or less. The Agency interprets this provision as applying only to sources for which specific regulations do not apply. EPA, however, argues that this provision applies to any source, and has cited the Four Corners Units 4 and 5 for violations. At a 1978 meeting the Board amended the smoke regulation to make the EIA's interpretation explicit.

Another regulation is concerned with schedules of compliance. Many of the emission regulations have "effective dates" in them, dates at which the regulation goes into effect. The Agency may require any source subject to such a regulation to submit, prior to the effective date, a compliance schedule showing how compliance is to be achieved by the effective date.

The "Source Surveillance" regulation gives the Agency the authority to require sources to keep records of emissions and submit reports on a regular basis. This clause is the regulatory basis of self-reporting requirements, and also, in the Agency's view, it gives the Agency the authority to order plants to perform source tests and report on the re-sults. This reading of the regulation has been disputed by at least one source, and remains subject to some question. In any case, the EIA has not imposed self-reporting requirements on very many sources. Natural gas plants, refineries and smelters must report sulfur emissions quarterly. For several years Arizona Public Service was requested to conduct monthly source tests on Four Corners Units 4 and 5 to measure particulate emissions. Beyond that, many plants of various kinds have been ordered to conduct source tests and report on the results, often after finding the source in violation.

Finally, there is a provision to allow for excess emissions during startups, shutdowns, and upsets. In these cases an operator must notify the department within 24 hours that a problem exists and outline what he

intends to do about it. Repairs must be made with maximum effort
(including shutdown or use of nonshift labor) and emissions must be kept
as low as possible. Excess emissions during periods of scheduled maintenance are also allowed, provided the EIA is notifed at least 24 hours in
advance.

Chapter 4

REGULATORY RELIEF: AMENDMENTS, VARIANCES AND
ASSURANCES OF DISCONTINUANCE, AND UPSET WAIVERS

A plant has several strategies available for avoiding pollution control expenditures. First, it can seek to have the regulation amended. This
approach is most ambitious but if successful most rewarding, because the
company can avoid all expenses connected with the difference between the
old and new emission standard. It can also avoid or postpone the installation of pollution control technology by securing a variance from the regulation. A variance may allow the source to postpone capital outlays and
avoid operating costs altogether for the duration of the variance (or it
may not, if the source must request the variance to take care of some unanticipated problem with equipment already installed). Finally, it can
fail to operate its control equipment in an effective manner, taking its
chances with enforcement. In this case what is avoided is--perhaps--a
portion of the level of operating expenses which would be necessary to
comply with the regulation.

Amending the Regulations: Two Case Studies

Six source-specific regulations have been amended at various times.
They are given in Table 4-1.

The amendments to the woodwaste burner regulations (Number 402) were
evidently not the result of an appeal from the industry. One tightened

Table 4-1. Amended Source-Specific Regulations in New Mexico.

Regulation Number	Type of Source (Pollutant	Date of Promulgation of Regulation	Date(s) of Promulation of Amended Regulation
402	Woodwaste burners (smoke)	1/23/70	6/18/70, 1/10/75
501	Asphalt processors (particulates)	1/23/70	6/26/71
504	Coal-burning equipment (particulates)	1/23/70	3/25/72. 12/13/74
506	Non-ferrous smelters (particulates)	1/10/72	12/10/76
602	Coal-burning equipment (sulfur)	3/25/72	12/13/74
622	Petroleum refining (sulfur)	6/14/74	7/8/77

the regulation, while the other contained provisions which tightened and loosened it. In one, the opacity limit was lowered from 40 percent to 20 percent. Although the other one allowed a longer startup period, it also added a temperature clause for night burning. Likewise the asphalt processing amendment was not particularly favorable to the industry, the most important change being the addition of a fugitive dust clause.

The other four amendments came about as a result of industry efforts to have their respective regulations weakened. At the behest of a small petroleum refinery in the state, the emission regulation for small refineries was changed from a certain percentage control efficiency to a maximum emission of 5 tons per day.

The particulate regulation for existing copper smelters was originally
0.03 gr/scf[1] for each stack, with an effective date of July 31, 1974. By
mid-1976 Kennecott had achieved compliance on the reverberatory feed dryer,
but not on the reverberatory furnace, acid plant and fire refining furnace.
Non-compliance in the period from 1974 to 1976 was excused through a variance
executed in July, 1974. Later in 1974 Kennecott sought an amendment to the
regulation, but the EIB refused to amend it at that time, and instead gave Kennecott a variance for two more years, until June 1, 1978. From Kennecott's point of
view the difficulty with this solution was that the company was still faced
with meeting the original requirements in two years, which meant that they
would soon have to begin investing in equipment which would enable them to
meet the regulation.

Therefore, Kennecott went before the Board one more time and was
successful. On December 10, 1976, the regulation was amended to allow the
emission rates then being achieved by Kennecott: 1,860 lbs/hr on the
reverberatory furnace, 0.06 gr/scf from the acid plant, 0.03 gr/scf on the
reverberatory feed dryer, and 20 percent opacity on the fire refining
furnace. The Board did leave in a requirement to meet the 0.03 gr/scr
standard by July 1, 1978. However, this provision did not have much
meaning, because it also promised by regulation that hearings would be
held within two years to reopen the question of appropriate permanent
emission regulations for the facility, and until the hearing was held the
agency was forbidden to require a compliance schedule.

---

[1] gr/scf = grains (1 grain = 1/7000 pound) per cubic foot of stack
gas, adjusted to standard temperature and pressure, i.e., this is a
measure of the concentration of particulates in the stack emissions.

The benefit of this amendment to Kennecott was considerable. At the very worst, it postponed for at least two years any investment in particulate control, and it offered the promise of even greater benefits if the Board decided to weaken the regulation in the 1978 hearing.

When the hearing finally took place, however, the Board kept the 0.03 gr/scf requirement for particulate emissions and raised the SO_2 removal requirements from 60 to 90 percent, with a compliance date of 1983.

Arizona Public Service (APS), the operator of the Four Corners Power Plant, has been if anything even more successful than Kennecott in securing favorable amendments to the regulations. APS was responsible for the amendments to 504 (the particulate regulation) and 602 (the sulfur regulation).

Regulation 504 was first promulgated on January 23, 1970, included among a large number of regulations issued on that day. These regulations were New Mexico's first state air quality regulations. The regulations for the Four Corners Plant were set at that time to be 0.135 lbs/10^6 Btu input for Units 1-3, and 0.5 lbs/10^6 Btu input for Units 4 and 5, with an effective date of January 1, 1972. For Units 4 and 5 this regulation reflected the design efficiency of the electrostatic precipitators. In fact, the precipitators had never achieved this level of performance, and a source test in mid-1972 indicated emissions of 2.5 lbs/10^6 Btu input on one unit and 5.6 lbs/10^6 Btu input on the other. Over the next six months the company, with some prodding by the EIA, was able to bring both units almost into compliance with the regulation. As of August 1978, neither unit has achieved full compliance, with about a quarter of all source tests showing

emissions in excess of 0.5 lbs/10^6 Btu input.

Shortly afterwards the federal Clean Air Amendments of 1970 were

passed, and New Mexico, along with the other states, had to prepare a state

implementation plan (SIP). In preparing the SIP it became evident that the

existing 504 would not be sufficient to meet federal ambient air quality

standards, and in August 1971 the Agency proposed to the Board an amended

504. Also proposed were Regulations 602 and 603, for control of SO_2 and

NO_x emissions from coal-fired power plants.

On March 25, 1972, the Board adopted these proposed regulations. The

amendment to 504 required that all units of the Four Corners plant achieve

an emission rate of 0.05 lbs/10^6 Btu input for total particulates by Decem-

ber 31, 1974, a tenfold improvement, with an additional limit of 0.02

lbs/10^6 Btu input for particulates smaller than two microns in diameter.

The SO_2 regulation was set at 1.0 lb/10^6 Btu input for existing plants and

0.34 lbs/10^6 Btu input for new plants. As it applied to the Four Corners

plant the limit for existing plants represented approximately 35 percent

removal of SO_2, about what would be incidentally achieved by the operation

of a high-efficiency particulate scrubber, which the company had indicated

it was going to install to meet the new 504. Thus, through prior negotia-

tion, the company had managed to tie together the achievement of particu-

late and SO_2 regulations, and would apprently not have to install a

separate flue gas desulfurization system on Units 4 and 5.

This solution was prevented by EPA, which in July 1972 found

that the SIP for New Mexico was not adequate with respect to SO_2 emissions,

and remanded 602 to the state. EPA based its decision on a high terrain

pollutant dispersion model built by the National Oceanographic and Atmos-
pheric Administration, which predicted that the proposed regulation would
allow ambient standards occasionally to be violated in the vicinity of the
plant (the model used by the state predicted no such violations). EPA
suggested 70 percent control, and the company filed a tentative compliance
schedule to meet this standard and the particulate standard by July 1,
1977.[2]

In July 1974 the EIA proposed an amended 602 which would require 65
percent control of SO_2 on Units 1-3 by mid-1977, 85 percent control on
Units 4 and 5 by 1977, and 90 percent control on Units 4 and 5 by 1979.
According to the subsequent court case, the Agency's justification was a
desire to allow additional siting of coal-burning facilities in the vicinity of the Four Corners Plant. By this time, San Juan Unit 2 was on line
and Unit 1 was well under construction less than ten miles from the Four
Corners Plant. In addition, El Paso Natural Gas Company and WESCO, a
consortium of pipeline companies, had each announced plans to build large
synthetic gas plants less than twenty miles south of the Four Corners
plant. If the limit suggested by EPA was just enough to prevent ambient
standards from being exceeded by the operation of the Four Corners plant,
then surely the construction of these facilities would cause violations
when their emissions were introduced into the NOAA model. A hearing before

---

[2]The particulate regulation called for compliance by July 1, 1974,
but at its discretion the EIA could allow in a compliance schedule an
extension of the effective date.

the Board was held and on December 13, 1974 the Board announced that it had

accepted the sulfur oxide regulation proposed by EIA. At the same time,

the effective date of compliance with 504D, the 0.05 $1g/10^6$ Btu particu-

late standard, was postponed from December 31, 1974 to December 31, 1977.

This change was made to allow the company to install equipment which would

achieve joint control of particulates and SO_2.

In 1975 APS appealed the 85 and 90 percent removal requirements for

Units 4 and 5 to the New Mexico Supreme Court, arguing that the regulations

were not required to meet the ambient standards, and therefore an unreason-

able use of police power. The Court agreed, and in April 1976 overturned

the regulation.[3] The Court ruled that the EIA's legislative charter in-

cluded only the improvement of environmental quality, not the promotion of

regional growth.

In September 1976, the EIA began negotiations with the company to

develop a sulfur regulation that both could live with. Agreement was

reached in December, and a new regulation was jointly proposed to the

Board. The new sulfur control regulation called for an overall 60 percent

control of SO_2 at the plant (which meant it could be achieved by greater

than 60 percent control on some units and less than 60 percent on others,

as long as 60 percent was obtained overall), to be achieved by the end of

1982. At the same time, compliance with the 0.05 $1bs/10^6$ Btu particulate

standard was to be postponed to the end of 1982 as well.

---

[3]Public Service Company of New Mexico, et al. v. Environmental
Improvement Board, 8 ERC 1899 (1976).

At the same time a coalition of citizen groups, including New Mexico Citizens for Clean Air and Water, the Navajo Tribal Council, and the Sierra Club, proposed an alternative regulation that would require 90 percent control by 1982. A hearing was held in August 1977, after which the Board ratified a compromise: 67.5 percent control of SO_2. In addition, ambient monitors would be set up in the vicinity of the plant--including several on high terrain nearby. If these monitors showed ambient violations, the plant would have to remove up to 78 percent of the SO_2. Shortly thereafter the compromise was amended, when APS agreed to achieve 72 percent control instead of 67.5, in return for an extension of two years to the compliance schedule.

Thus, APS will have to install additional SO_2 and particulate control equipment at the Four Corners Plant no sooner than eight years after it was originally required. For SO_2 this was due to the company's success in challenging the regulation in court; for particulates it was successful because it convinced the Environmental Improvement Board of the cost advantages of simultaneous control of particulates with SO_2.

An amended air quality regulation is, of course, an amendment to the state SIP, and therefore it must be approved by EPA. However, neither the amended particulate regulation for coal-combustion nor the amended smelter regulation was ever approved by EPA. The matter had not been pressed by EPA Region VI until the summer of 1977, when EPA found Kennecott in probable violation of the original 506 (0.03 gr/scf for each stack) and ordered source tests. These tests were conducted in March 1978, and the plant was found in violation of the regulation for the reverberatory furnace, the reverberatory dryer, the converter stack, and the refining furnace. In June the plant was notified that compliance was to be achieved

by 30 days or a federal compliance order would be issued, with possible

civil or criminal action. However, this action was not pursued. When the

Board promulgated amended smelter regulations the following year, EPA

approved the amended regulations as part of the state implementation plan.

A similar series of events has ensued with respect to the Four Corners

plant. In January 1978 EPA made opacity readings on Unit 4 and 5, finding

results of 92 and 90 percent, respectively. EPA then notified APS that

the plant was in violation of Regulation 401, which limits a source of

smoke to 20 percent opacity,[4] and probably also in violation of 504 (the

version of 504 recognized by EPA required emissions no greater than 0.05

$lbs/10^6$ Btu), and ordered a stack test. Joint control is not accepted by

EPA as a rationale for exempting the plant from more stringent particulate

control, and the plant has been ordered to come into compliance with the

$0.05 \ lbs/10^6$ Btu standard by July 1, 1979. Again, however, EPA failed to

pursue sanctions against the company, apparently deciding to accept

the outcome of the Board hearing the previous August. Although EPA has

not yet formally accepted the agreement, it probably will, since the

agreement has the support of the EIA, the company, and environmental

groups in the state.

---

[4]The position of the EIA is that 401 does not apply to any source
whose emissions are limited by a more specific regulation, such as coal
combustion. Probably this regulation will soon be amended to make this
explicit.

Variances and Assurances of Discontinuance

The variance and the assurance of discontinuance are two devices used in New Mexico to provide some flexibility to the enforcement process by allowing sources to be out of compliance without threat of penalty. A variance suspends a regulation as it applies to a particular source, as long as certain conditions are met. The assurance of discontinuance is granted in some situations to plants with sources which have been found to be in violation of a regulation, and is something like a consent decree.[5] The plant agrees that the Board considers it in violation, and promises to be in compliance by some future date. In return, the Board promises to refrain from any further enforcement actions until after the expiration of the assurance. Both are important elements of New Mexico's voluntary compliance approach, to be described in greater detail in the next chapter.

According to the New Mexico Air Quality Control Act, a variance can be granted if there is no practicable way of meeting the regulation, or if enforcement of the regulation will result in a taking of private property or impose an undue economic burden on the source. An example of the former is the variance granted to Arizona Public Service for NO_x emissions at Four Corners Unit 3. APS convinced the Board that the company did not know how to meet the emission standard on Unit 3, and was granted an indefinite variance until the technology becomes available. For almost all variances, however, the economic hardship argument is the formal rationale

---

[5]Variances and assurances of discontinuance have been negotiated with both officers of firms and with individual plant managers, the latter especially for plants with out-of-state ownership. For ease of exposition the ensuing text will refer to the entity seeking the variance or assurance of discontinuance as the "plant."

for granting the variance. In this case the variance must specify a certain date by which compliance is to be achieved, together with a compliance
schedule. Finally, under no circumstances will the variance result in a
"condition injurious to health or safety," although this does not mean that
the primary ambient air quality standards must be met at all times. Several
variances have been granted where it was recognized that ambient standards
would likely be violated on occasion.

Seventy percent of the variances which have been approved have been
granted before the effective date of the relevant regulation. The remainder have been granted to plants experiencing difficulties with their control equipment, or plants which need to expand their production capacity.
All assurances of discontinuance, of course, are entered into after a
violation has been cited.

The procedure for applying for a variance is illustrated in Figure
4.1. As shown, a plant submits a variance petititon to the Board. The
application is studied by the EIA, which makes a recommendation to the
Board. If the recommendation returned by EIA is favorable, the Board
schedules a public hearing. If the findings result in a recommendation
that is unfavorable, then the Board notifies the plant, which
may then request a hearing. Otherwise, the variance is denied. At the
hearing, the Board considers evidence presented by the applicant, the EIA,
and any other interested parties and makes its decision. If denied the
variance, the plant may appeal to the state court of appeals.

Since 1970 the Board has granted 23 variances and denied ten others.
All but two of the denials have been applications for a variance from the
open burning regulation, and most of these were made in 1970 or 1971.

Figure 4.1

New Mexico Variance Application Procedure

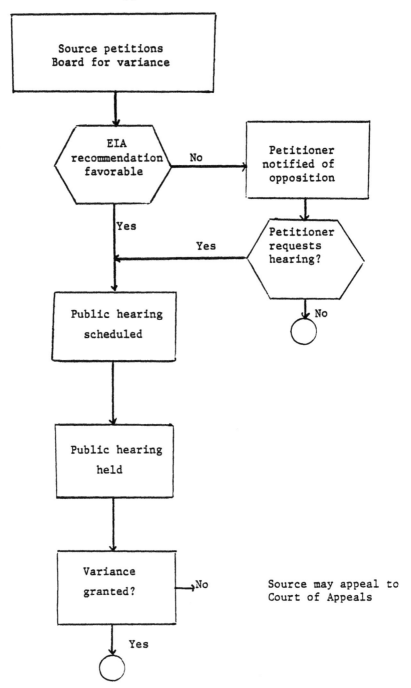

Five applications came from salvage yards wishing to burn junked automo-

biles, and two others came from local governments wishing to incinerate

solid waste. Thus, early on a precedent was established that the open

burning regulation would be strictly interpreted. This reflected the posi-

tion that there was usually an alternative means of disposal in which

burning was avoided. As for the other two variances which were denied,

one came from a paving company and one came from a sawmill. Of the

successful applications for variances, all but two came from industrial

sources, which ordinarily did not have an alternative which eliminated

emissions.[6] But although all but two variances for industrial plants have

been approved, the Board regularly imposes conditions in addition to a time

limit, such as a limit on emissions during the variance period, compliance

with regulations on other pollutants, or no ambient violations. No plant

has yet appealed a denied variance.

While satisfying the same function, an assurance of discontinuance

differs from a variance in several ways. First, it is, by definition,

granted after a violation has taken place. The plant must sign a statement

agreeing that it has violated the regulation or, more commonly, that it is

"deemed in violation" by the EIA. Assurances of discontinuance must also

---

[6]For all industrial sources except woodwaste burners, the airborne
emissions are intrinsic to the production process. Woodwaste burning is
a means of disposal of solid waste. However, unlike the burning of junked
cars or municipal refuse, the burning of woodwaste could be avoided only
with great difficulty. An average-size sawmill creates an enormous amount
of waste, which may burn spontaneously if stockpiled. Utilization is an
alternative, but this is only now becoming economically feasible on a
large scale in the Southwest. (In the Pacific Northwest woodwaste has
been used for years for fuel and for paper production.)

designate a date when compliance will be achieved; moreover, the average

duration of an assurance is considerably less than that of a variance (6.5

months compared to 21 months for those variances with termination dates).

This in part reflects the fact that assurances are intended to allow a plant

time to remedy some technical problem, rather than mitigate the economic

impacts of the regulation. Finally, unlike a variance, an assurance does

not require a public hearing. The procedure for entering into an assurance

is therefore much simpler. Almost invariably, assurances of discontinuance

are the result of negotiation between the EIA and the source. In fact, the

Agency, when informing the plant of a violation, occasionally will offer to

support an assurance if the plant wishes to apply for one. Typically, an

initial version of the assurance is drafted by the plant and sent to the

agency for its concurrence. The agency makes whatever changes it feels

are necessary and submits it to the Board.

The behavior of plants with respect to variances and assurances of

discontinuance appear to be a characteristic of the industrial category.

Table 4-2 gives the number of firms to be granted variances or assurances

of discontinuance by industrial category, the total number of variances

and assurances granted those plants, and the fraction of the whole cate-

gory represented by these plants. At one time or another virtually all

the coal-fired steam generators, smelters, and nonmetallic mineral pro-

cessing plants in the state have had variances or assurances. At the

other extreme, no gas-fired utility plants, sulfuric acid plants and very

few asphalt processing plants have been granted them. In the middle, about

half the woodwaste burners, petroleum refiners and natural gas processors

with sulfur plants have at one time or another sought relief. Part of

this pattern is easy to explain. Gas-fired utility plants and sulfuric

Table 4-2. Variances and Assurances of Discontinuance by Industrial Category.
(exclusive of Bernalillo County)

Category	Total Number of Plants	Number of Plants Receiving Variances[a]	Percent of Total	Number of Plants Receiving Assurances of Discontinuance[a]	Percent of Total
Natural gas processing	8[b]	4	50	2	25
Petroleum refining	7	4	57	0	0
Utility steam generators					
Coal-fired	3	3	100	2	67
Gas-fired	8	0	0	0	0
Oil-fired	2	0	0	0	0
Asphalt processing	27	0	0	0	0
Woodwaste burning	12[c]	0	0	7	70
Sulfuric acid plants	2	0	0	0	0
Nonferrous smelters	2	2	100	0	0
Nonmetallic minerals	8	1	13	6	75

Notes:

a. Some plants received more than one variance or assurance of discontinuance, so the number of plats is different from the total variances or assurances granted.

b. Excludes plants not subject to sulfur regulations.

c. Includes two woodwaste burners that went out of business.

48

acid plants did not have to install additional equipment to meet the regulations, and in general compliance with air quality regulations is not
difficult for them. Several economic conditions may have had much to do
with the relief requested by woodwaste burners and smelters. Lumber production is a declining industry in the state, and at least one assurance
was granted until a certain date or until the financial health of the
company improved, whichever came first. This company was one of at least
three mills which have gone out of business since the early seventies. The
copper industry, after a prosperous period in the late sixties (during
which time the Phelps-Dodge plant, a "new source," was planned) fell into
a depression which only recently has begun to abate. When we look more
closely at terms of the variances and assurances granted, together with
the performance of the firms operating under them, some other interesting
facts emerge. Again, the discussion is best taken up by industrial category.

Natural gas plants. Half the natural gas plants with sulfur recovery
units have gotten variances to help meet the sulfur regulation. Typically
these variances were for short periods, at most a year. Each of these
plants was able to achieve compliance promptly at the termination of the
variance. One plant sought a variance for NO_x emissions. Although there
is no NO_x regulation for gas plants, emissions from this plant caused a
violation of ambient standards. This firm requested a variance when it was
denied a permit to expand its operation on the basis of noncompliance with
the ambient standard.

Petroleum refineries. New Mexico has a vapor recovery regulation for
tank farms, and two sources sought relief from this regulation.

These variances are still in effect. Two other refineries sought relief from the carbon monoxide regulation. CO is produced in these plants in catalytic cracking units; when these units were expanded recently it caused the regulation to be violated. One refinery is now back in compliance with the regulation. Its response was to switch to a different catalyst which does not produce as much CO, and also to flare the CO that is produced, converting it to CO_2. No variance was requested for sulfur recovery, probably the most expensive pollution control task at a refinery, possibly because of the small refinery exemption discussed earlier.

Utility Steam Generators. For the Four Corners Plant, Arizona Public Service requested variance for excessive NO_x emissions. A variance negotiated late in 1974 postponed compliance with the NO_x emission regulation until the end of 1977 for Units 1-3. When it expired only Unit 1 was in compliance. For Unit 2, the company requested an assurance of discontinuance until the end of 1978. On Unit 3 the company claimed that the particular boiler design prevented achievement of the standard with known technology, and the EIA agreed to an indefinite variance. All the while, it was understood by both APS and the EIA that Units 4 and 5 were in compliance on NO_x; in fact, the three-year variance on Units 1-3 was granted on condition that Units 4 and 5 would be in compliance. Both these units were equipped with continuous NO_x monitors, which showed the units to be marginally in compliance. (On both units a low-NO_x mode of operation had been adopted, the main feature of which was taking certain burners out of service to reduce boiler temperature.) In late 1976 the company did request an assurance of discontinuance for a short period in order to investigate the cause of poor combustion at both units. One of the outcomes of this series of experiments was the discovery that the NO_x monitors at the plant were reporting NO_x concentrations about 25 to 30 percent below the true values. No one knows how long the reporting malfunction had existed, and therefore how long Units 4 and 5 had been out of compliance.

As discussed earlier, there has been no need for APS to seek variances from particulate and SO_2 emissions, because of its success in having the regulations changed. By fighting regulations on particulate and SO_2 and by securing variances on NO_x, the company has successfully avoided almost all emission regulations. At present, the only state emission regulations which are applicable are the NO_x regulation on Unit 1 and particulate regulations which, especially on Units 4 and 5, are extremely lenient.

With respect to stringency of the regulations, the contrast between Four Corners and San Juan, the other large coal-fired power plant in New Mexico, could hardly be more complete. This is clear from Table 3.2. To meet the more stringent SO_2 standards Public Service Company of New Mexico has been granted variances on Unit 2 and the yet-to-be-built Unit 3. The variance on Unit 2 was necessary because a boiler explosion had put the plant out of service for a lengthy period and postponed the installation of pollution control equipment.

Woodwaste burners. About half the woodwaste burners in New Mexico have entered into assurances of discontinuance at one time or another, but most of the assurances have been granted to two companies: Blevins Lumber and Duke City Lumber. These firms have been granted series of assurances while they sought a means to utilize their wastes rather

than burn them. Since the regulation became effective the EIA has sought
to replace incineration with utilization, and as long as these companies
were actively seeking such a solution they were supported by the EIA.
Blevins actively promoted the use of pyrolysis; their plan was to
produce charcoal out of the woodwaste. However, after years of trying
the company was never able to make the process work satisfactorily, and
finally gave up. During the period from mid-1975 to mid-1977 Duke City
also investigated pyrolysis, but abandoned the idea just as Blevins did.
Instead, Duke City has been able to find a user for the waste from both
its Cuba and Espanola mills for the manufacture of particle board.
This manufacturer may take as much or as little of the woodwaste as it
likes, so on occasion one will find that the burners at Cuba and
Espanola must be operated.

It would seem reasonable that the Agency's promotion of woodwaste
utilization would be entirely consistent with its primary goal of
improving air quality, but, surprisingly enough, it is not, at least
not completely. As noted, the primary cause of excessive smoke is
operation at low temperature. In turn, low temperatures usually result
from either insufficient fuel relative to the size of the burner or
the burning of wet or green fuel. Both are problems which may be
exacerbated by partial woodwaste utilization. The Duke City Espanola
Mill is a case in point. The burner at the mill is sized for full
incineration, and now that part of the waste is being utilized it is
more difficult to maintain adequate temperatures. While an appropriate match between fuel feed rate and burner size could be maintained
if the unused waste were stockpiled and processed in a batch mode,

the mill is already equipped with an automatic feed into the burner, and batch operation would entail considerable extra handling. An additional difficulty is caused by the fact that the waste which cannot be utilized at present is precisely that which is most difficult to burn: green sawdust (the user of the waste is currently undergoing tests to determine the maximum amount of green sawdust they can incorporate into their particle board without an unacceptable weakening of the product).[5]

Nonmetallic minerals. The experience at several of the non-metallic mineral processing plants in the state dramatically illustrates the point that the date at which Agency and firm agree or predict that compliance will be achieved is not necessarily the date that compliance is in fact achieved. According to assurances of discontinuance signed by Grefco, Johns Manville and J.H. Rhodes, Inc., those firms were to achieve compliance at their New Mexico plants by late 1972 or early 1973. However, actual compliance was not achieved consistently at any of these plants until four years later. In the interim, all three plants were usually operating in violation of emission regulations without any formal sanction from the Board. The reason for this was not that in 1973 the EIA decided to grant each plant a de facto variance of four years. On the contrary, the long delays were unanticipated by the agency and each resulted from a series of shorter delays—to order

---

[5] Conversely, the woodwaste burner with the best opacity control record in the state has achieved this enviable record by incinerating a salable product. White Sands Forest Products of Alamorgordo claims to have a buyer willing to pay $2,000 per month for its dry planer shavings, but which are used instead to maintain high (1200°F) furnace temperatures.

equipment, install it, repair it--which the agency permitted in its observance of the voluntary compliance requirement. The experiences at these plants is discussed more fully in Chapter 5.

Explaining the patterns. Evidently, sources in some categories tended to seek variances, sources in other categories tended to seek assurances of discontinuance, and sources in still others sought neither. The reason for this pattern is not entirely clear, but in Chapter 7 some hypotheses attempting to explain these results are suggested.

Upsets, Breakdowns, or Scheduled Maintenance

New Mexico's air quality regulations also provide for regulatory relief in the extreme short term. Whenever an upset condition or scheduled maintenance is the cause of excess emissions, it shall not be considered a violation of the regulations as long as certain conditions are met. These conditions require first that the Agency be notified as quickly as possible following an upset and in advance for scheduled maintenance. Second, the repairs must be made with "maximum reasonable effort," including use of overtime if necessary. Third, emissions should be held to a minimum during the period of the breakdown, and the public health or safety should not be impaired. Finally, the Agency may revoke this waiver if the breakdowns occur with such frequency that careless operation is suggested.

In terms of emissions, probably the most important case in New Mexico for which the upset clause is invoked is for startup of coal-fired power plants equipped with electrostatic precipitators. During startup the electrostatic precipitator must be bypassed until the flue

gas temperature exceeds the dew point. Otherwise, condensation in the
precipitator may cause arcing, leading to equipment damage.

In this section we examine the record of startup on Units 4 and 5
of the Four Corners Plant, estimating the magnitude of particulate
emissions resulting therefrom. With a generating capacity of 800
megawatts each, Four Corners Units 4 and 5 are two of the largest
coal-fired steam generators in the Southwest. They are also two of
the earliest, having been installed respectively in 1969 and 1970.
Both units burn coal from the nearby Navajo Mine, which has a heating
value of about 9000 Btu/lb and an ash content of about 22 percent by
weight. The emission control system of each unit is a cyclone capable
of removing about 70 percent of the flyash in the stack gas, followed
by an electrostatic precipitator capable of achieving the emission
standard of 0.5 lbs/10^6 Btu input when operating properly.

The magnitude of annual startup emissions depends on the frequency
of startup and the total emissions during each startup period. Though
primarily base loaded, Units 4 and 5 frequently suffer shutdowns,
which of course means frequent startups. Table 4-3 presents the
startup frequency experienced on Units 4 and 5 during the years 1972 to
1975; as shown, each unit averaged 19 startups per year.

For each startup a report must be sent to the EIA giving the
duration of the period for which particulate emissions were uncontrolled
and the plant's estimate of total emissions during the period. These
estimates are calculated not from source tests made during startup,
but from emission factors relating the known operating rates during
each startup with uncontrolled emissions at those rates. A coal-fired

Table 4-3. Frequency of Startup on Units 4 and 5 Four Corners Power
           Plant[a]

Year	Unit 4	Unit 5
1972	22	17
1973	20	18
1974	18	18
1975	16	23
AVERAGE	19	19

[a]Each unit is 800 MW capacity.

boiler is brought up to speed in stages, and at each stage certain
systems must be checked before proceeding to the next stage. Hence,
due to unanticipated problems which often arise, the duration of the
startup period and the emission rate vary considerably.

The startup reports for 1972 to 1975 show that the period of
uncontrolled emissions varied between 2 and 69 hours, with a mean
value of 16 hours. The emission rate for a sample of 39 startups
varied between 2.5 and 11.9 tons of particulates per hour, with a
mean of 7.2. By comparison, the emission rate when a unit is operating
at capacity and in compliance with the particulate emission regulation
(0.5 lbs. per 10^6 Btu input) is 2.0 tons per hour. Annual startup
emissions are estimated to be 2,300 tons per unit, compared to the
total annual particulate missions during normal operation from

these units of about 21,000 tons per unit, so that startup emissions are about 11 percent of "normal emissions." If these units should have to meet the more stringent requirement of 0.05 lbs per 10^6 Btu, then startup emissions would likely exceed normal emissions.[6] The actual impact on air quality of startup emissions is probably more severe than the estimates of relative quantities would indicate, because they are concentrated in time.

New Mexico is fairly typical in its tolerance of startup emissions. But these observations suggest that startups of coal-fired boilers may be a significant source of particulate emissions. If so, then it is worth considering whether there are alternatives to continued uncontrolled emissions during startup. One possibility is to establish incentives to minimize the number of startups. However, steam plant operators already have a very strong incentive to minimize startups. Every startup is preceded by a shutdown, and every hour of downtime for a unit the size of Four Corners 4 or 5 costs its operator about $15,000 in replacement power. Another possibility is to require the burning of natural gas during the startup period, and it will now be shown that under some circumstances this alternative may be cost-effective.

The marginal cost of eliminating startup emissions, expressed in dollars per pound of particulate emissions reduced, is

$$MC = \frac{D}{FA} ,$$

---

[6]This might not occur if the Four Corners Plant meets the more stringent requirement by installing scrubbers, because scrubbers may not need to be bypassed during startup.

where

> D   denotes the difference in cost between coal and
> natural gas, in dollars per 10^6 Btu,
>
> A   denotes the ash content of the coal, in pounds
> per 10^6 Btu, and
>
> F   denotes the fraction of the ash that escapes the
> stack during startup.

Assuming a value of D of $2.00 per million Btu, the marginal cost of reducing startup particulate emissions from Unites 4 and 5 is about 25¢ per pound of particulates reduced. This figure is lower than it would be for other plants because Navajo coal has an unusually high ash content (22 percent). For coal with the same heating value containing ten percent ash, the cost of avoiding startup emissions would be about 60¢ per pound. Nonetheless, even the lower value is somewhat greater than the marginal cost of reducing particulate emissions at the Four Corners Plant at the existing level of control.

As the removal efficiency of pollution control devices approaches unity, the marginal cost increases rapidly. At some point, the marginal cost of ever more stringent control of particulate emissions during normal operations will exceed the cost of controlling particulates during startup, and it becomes cost-effective to control startup emissions. It may well be that that point has been passed.

The Clean Air Amendments of 1970 required EPA to set new source performance standards (NSPS) for stationary sources in a number of industrial categories. For coal-fired boilers, the NSPS promulgated in 1972 set an emission limitation of 0.1 pounds of particulate emission per 10^6 Btu input.

The statute also required EPA to review and if necessary revise these standards periodically, as advances in technology made more stringent standards feasible. Recently, EPA completed such a review and in 1979 revised the NSPS for utility boilers, limiting particulate emissions to 0.03 pounds per 10^6 Btu input. This standard would apply to any source for which construction began after the promulgation of the new standards.

Based on cost data submitted by EPA in support of the proposed regulation, it is estimated that the difference in cost between the old NSPS and the new NSPS for particulate emissions is 0.4 mills/kwh for low-sulfur coal containing ten percent ash.[7] This corresponds to about 65¢ per pound of particulates removed, compared to 60¢ per pound for reducing startup emissions.

This exercise shows that the cost of controlling startup emissions of particulates is comparable to the marginal cost of the new source performance standards for particulates. In any particular situation the actual comparison will be quite site specific, and will, in any case, depend on the actual cost difference between coal and natural gas. What the above calculation suggests is that control of startup emissions is something that should be considered in regulating new steam generating plants. This is all the more true when one considers the impact on air quality of the concentration of startup emissions in a short period of time.

---

[7]U.S.E.P.A., "Electric Utility Steam Generating Units: Background Information for Proposed Particulate Matter Emission Standards," Research Triangle Park, N.C., EPA-450/2-78-006a, July 1978. For details on the calculation of the marginal cost, see Winston Harrington, "Effectiveness and Cost of Air Pollution Control for Coal-fired Electric Power Plants in the West," Unpublished manuscript, Resources for the Future, June 1978.

CHAPTER 5

ENFORCEMENT PRACTICES

General Issues

Previous chapters have discussed governmental organization for air

quality management, especially at the state level; the sources and the

regulations which apply to them have been introduced; and the various ways

in which a source can secure relief from a regulation have been described.

Now we turn to the practices and techniques used by the EIA to secure

compliance with the regulations.

We may think of the enforcement problem as having two aspects. First,

the source must be made to acquire at least the potential for complying

with the regulation. This condition may be satisfied in several ways.

End-of-pipe control equipment may be installed. The plant might also

change the production process so as to reduce the generation of regulated

pollutants. Or it might promise to use raw materials that create pollu-

tants in smaller amounts. Second, this potential for compliance must be

translated into performance. If new equipment is installed, it must be

operated in an effective manner, so that emissions are maintained within

the regulation. If a different raw material is used, it must be used

continuously. This distinction is unfortunately not as unambiguous as it

might appear; there is a gray area, a situation which is hard to classify into one of these categories. This occurs when equipment meeting the regulation is really not adequate to the task, when the standard can be met only under ideal conditions. As will be discussed below, there is an important case of this kind in New Mexico: two perlite mills installed equipment which could not consistently comply with the regulation.

Nonetheless, the distinction made here between initial compliance and continuous or routine compliance is a useful one because it corresponds to the distinction between construction and operation. Indeed, any operating permit system contemplates this two-step approach to enforcement, where obtaining the permit is analogous to initial compliance. Of course, it is continuous compliance, not initial compliance, which is important as far as air quality is concerned, but initial compliance is a necessary condition for continuous compliance. Much of the effort of an air pollution control agency will be directed at ensuring that stationary sources can meet the regulation at least once.

Another reason to think of two kinds of compliance is that a plant's incentives for noncompliance depend in part on whether any equipment has been installed. If, before the investment in control equipment, a plant is successful in postponing for one year the date of compliance with a regulation, it saves a year's worth of operating expenses plus the value of postponing the investment for one year. Inasmuch as end-of-pipe pollution control equipment tends to be rather intensive (for an electrostatic precipitator, for example, the annualized capital charge may be more than ten times the annual operating cost), the virtues of postponement can be substantial, at least from the firm's viewpoint. On the other hand, if the plant is out of compliance for a one-year period after the equipment has been installed, no capital charge is saved. Further, the operating expense is avoided entirely only if the equipment is not operated at all. If the equipment is operated, but not so as to be in compliance, the expense avoided is the difference between actual costs and what costs would have had to have been to have achieved compliance. Finally, if the pollution control system recovers a salable product, the net operating costs will be reduced by the value of the recovered product, and may even be negative. In that case the plant has a positive incentive for continuous compliance.

The difference in incentives may be even more pronounced for those dischargers who have chosen to meet the standard by a method other than end-of-pipe treatment. If a process change has been chosen, a failure to maintain continuous compliance means a failure to operate process equipment properly, which may cause a loss of output.

It should not be inferred from these comments that plants fail to achieve continuous compliance necessarily as a result of a willful disregard of the regulations (although that can happen and is certainly sufficient for noncompliance). More likely, violations result from stochastic events, which can occur either in a source's production process or its abatement equipment. For example, an upset in the production process can generate more residuals than the abatement equipment can handle. Nonetheless, there are usually actions that can be taken, at some cost, to reduce the probability of such occurrences. These include the initial installation of inherently more reliable equipment, frequent preventive maintenance, and more careful operation. To the extent that such measures are available, plants can prevent emissions in excess of the regulations.

In any case, with two major exceptions and several minor ones the EIA has virtually completed the task of implementing initial compliance among the stationary sources in New Mexico. (The major exceptions are the Four Corners power plant and the Kennecott Copper Smelter. As far as the state is concerned, even these plants are in compliance. However, they are in compliance only by virtue of amended regulations which postpone the installation of additional pollution control at least until the 1980's.) Initial compliance was not accomplished immediately, as is evidenced from the large number of variances negotiated. Many plants were not brought into compliance for the first time until as much as four years after the effective date of the regulation. But the startup phase of enforcement in New Mexico is just about over now, and the enforcement effort in the future will consist largely of ensuring continuous compliance.

So, in this chapter we turn to a consideration of the enforcement of continuous compliance with New Mexico's air quality regulations. In general any enforcement program must have three features, and it is usually thought to be highly desirable to have a fourth. The three essentials are:

(i) for each source, the definition of what constitutes unacceptable behavior—a definition of compliance;

(ii) the ability to monitor sources to determine if emissions are in compliance; and

(iii) the power to implement measures which would induce the restoration of compliance in the event of a violation.

An enforcement program which lacked any of these elements would clearly be unable to affect source emissions at all.

The fourth feature is an ability to prevent emission violations from occurring in the first place. This is not, however, a necessary feature, and to see this one need only consider the example of a thermostat. A thermostat controls the operation of a heater, say, by continuous monitoring of temperature. When temperature becomes too high or too low, the thermostat has the ability to turn the furnace off or on, respectively. But it does not have the ability to give the heater a target temperature, and the heater certainly has no motive or ability to modulate its operation to achieve that temperature. Similarly, if an air quality control agency had perfect information about the emissions from sources under its control, together with the power to effect instantly a correction of any violation reported, nothing else would be needed to achieve compliance with the regulation. However, information on emissions is usually far from complete, and it inevitably takes time to bring a source back into compliance. Thus, to the extent that these ideal conditions are not met, an air quality agency needs to be able to prevent violations from occurring.[1]

Because plants often have strong economic incentives not to comply with emission regulations, they must be provided with at least equally

[1] In some enforcement situations it may be argued that lack of perfect information and perfect ability to correct violations are not the only reasons for the application of sanctions. In these situations the consequences of a violation are irreversible or otherwise so serious that remedial action alone is not sufficient.

strong incentives for compliance; otherwise the regulations will be disregarded. It is usually assumed that such incentives are provided by the threat of legal action, which not only forces the plants to commit legal resources but also may result in a fine or injunction.

If an air quality regulation is violated in New Mexico, authorities can seek an injunction or a fine not to exceed $1,000 per occurrence (each day of violation shall be considered a separate occurrence). However, legal actions can be sought only after a plant has been given the chance to comply "voluntarily" with the regulation. This is very much in contrast to the authority of mine safety inspectors or health department inspectors, who have the authority to shut down an operation on the spot. Of course, violations of mine safety or public health regulations may cause an immediate threat to the public. Such a threat is not likely with violations of air quality regulations. In any event, "voluntary compliance" has a very important place in New Mexico's enforcement program, and will be considered in some detail below.

In addition to formal legal action, there may be several other "informal" incentives which encourage compliance. For example, a plant found in violation may be ordered by the EIA to conduct a source test. This is not an insignificant expense, especially for small plants. In addition, the Agency could make things difficult for recalcitrant plants when new air quality permits are needed. This incentive may not be quite as important in New Mexico as in some other states because only a construction and not an operating permit is required. Nonetheless, any source wishing to expand would have to get a construction permit, and this requirement conceivably could act as an incentive for compliance.

Other potential incentives for compliance may be less under Agency control. Perhaps other units of state government could also make trouble for the source. For example, the state highway commission could avoid doing business with any paving contractor not meeting emission regulations. Use of this incentive would require the cooperation of other units of state government, which might require those units to violate some of their own policies or preferences. Another incentive not under the control of the Agency is the weight of public opinion. This incentive would operate toward better compliance among sources in populated areas than elsewhere.

Finally, in some situations it may be in a plant's self-interest to remain in continuous compliance with the regulation, especially after initial compliance has been forced. Generally, we can say that the magnitude of the incentive to comply depends on the characteristics of the control technology.

At one extreme, consider a source for which the operating and maintenance (O&M) costs are relatively high and for which the rate of emissions is very elastic with respect to O & M costs. A good example is a venturi scrubber, which is relatively inexpensive to install, but requires high energy and water inputs (the latter often a scarce commodity in New Mexico) to operate effectively. With equipment with these characteristics, cost minimizing behavior might be to cut back on O & M costs and allow excess emissions until a violation is discovered. Likewise, processes or control equipment which lead to a highly stochastic rate of emissions may give rise to a pattern of frequent violations. Even after a violation is discovered, a plant may well decide not to invest to improve the reliability of the process or equipment, and hope instead that on the next visit emissions will be in compliance.

On the other hand, if O & M expenses are relatively small--as with an electrostatic precipitator, for example--then the potential savings offered by noncompliance are also small, and may easily be outweighed by public relations or other considerations. Finally, there will be cases where effective O & M is of positive benefit to the firm. Two examples are (i) a source for which proper operation and maintenance prolongs the life of the control equipment, (if the incremental O & M costs are less than the savings in investment from postponing the replacement of equipment) and (ii) a source for which the control equipment captures a salable by-product capable of generating revenues sufficient at least to cover operating and maintenance expenses. In each of these cases effective source control probably needs no further spur from the state.

It would obviously be interesting and important to know which of these incentives, or perhaps others unmentioned, had the most effect on the behavior of plants. Unfortunately, that has not proved possible, mainly because the "or else" clauses implicit in all of these alleged incentives ("comply with this regulation or else ...") simply have not materialized, at least not observably. As will be discussed below, formal sanctions have been applied only in a handful of cases. Only one permit application has been denied, and in any case no attempt has been made to link past performance with new permits. Likewise, no other state agency has overtly taken action on behalf of air quality. Apparently, whatever compliance there has been has been caused by the threat of action rather than the action itself. This makes it very difficult to tell what the incentives really are.

One reason that such actions have been taken so rarely is the reliance on "voluntary compliance" mentioned earlier. Whenever a violation of a regulation is discovered, the EIA must, by statute, first give the offending source a chance to come back into compliance voluntarily. Though there is no statutory requirement as far as the other possible incentives are concerned, the EIA nonetheless give sources a chance to comply voluntarily before imposing any kind of sanction. Thus the philosophy of voluntary compliance pervades the Agency and exerts a profound effect on the way the EIA conducts its business.

Procedurally, voluntary compliance works in the following way. When, as a result of source test or inspection, a plant is found to be not in compliance, the department sends the owner or manager a letter informing him of that fact, and asks him to reply within two (or sometimes four) weeks and inform the agency what he plans to do to come back into compliance. The owner may respond in one of several ways.

First, he may reply that he was experiencing an upset condition, or there was some other unusual circumstance which caused his equipment to perform improperly. To this news the Agency may reschedule the inspection and informs the operator to submit an "801" Report in the future. (Pursuant to Air Quality Regulation 801, a plant is supposed to notify the EIA within 24 hours of any upset or startup condition which leads to excess emissions.)

Second, the owner may report that its control equipment was in need of repair, such as new filter bags needed, a new pump to be installed,

and the like. Generally, it will report that arrangements have been made
to get the equipment repaired, and gives the agency a date when it
expects to be back in compliance. If this date is not too far into the
future, and if the source has not been a persistent violator in the past,
the EIA will generally accept this solution and either schedule an inspection or source test for shortly after compliance has presumably been
achieved, or else order the plant to conduct a source test and report on
the results.

Often, however, the owner may not know why he is having problems
with his equipment, or if he does know the condition may not be presently
correctable. In that case, the agency and the source may have a meeting
and negotiate an assurance of discontinuance or a variance. As noted in
Chapter 4, each is a type of agreement between the agency and the source
in which the agency agrees not to enforce the regulation as long as the
source meets certain stipulations.

The final way in which an owner may respond is by defiance. He may
simply decline to answer the letter from the EIA notifying him of a violation, or he may deny the validity of the test result or even the
regulation itself. Defiance is rarely if ever the first response of a
source; usually it comes after a period of negotiation between the source
and the Agency which results in an impasse.

If, in the view of the Enforcement Section, the plant is not complying voluntarily, a letter is sent from the Legal Section of the EIA. This
letter informs the plant that if it cannot show evidence of progress
toward compliance within a certain time limit, then legal proceedings
will be initiated. The EIA will send this letter if it encounters

definance, of course. A threatening letter is also sent if it appears
that the source is dragging its feet or otherwise being less than forth-
right. If this letter does not produce results, then the EIA will
eventually resort to court action. The case is turned over to the State
Attorney General, who may decide to file a complaint. It is extremely
rare for a case to get this far without resolution, and the elapsed time
before this step is reached is measured in years, not months.

On the evidence, it would appear that the voluntary compliance
approach offers the second essential characteristic discussed above; that
is, once a violation has been discovered, the EIA has almost always been
successful in bringing the source back into compliance. Usually this
has been done simply by sending the owner a letter. Most of the remain-
ing cases were resolved through negotiation, with only a very few cases
in which legal action was threatened, and still fewer in which it was
actually used. But while a source can be returned to compliance, it is
apparent that the voluntary compliance approach does not provide any
incentives to a plant until after a violation has been discovered, because
a plant knows that it will have a chance to come back into compliance
before sanctions are imposed. Due to the fact that few violations are
discovered and fewer are corrected immediately, the lack of an incentive
may mean stationary sources in New Mexico are spending a lot of time in
violation.

In the next chapter, tentative estimates of the amount of time in
violation by source type will be made, based on the available surveillance
data.

However, in this study we will not be able to attribute any lack of compliance to the voluntary compliance approach. To warrant such a conclusion one must compare New Mexico's experience to the record in states where voluntary compliance is not relied upon.

In any event, many state officials in the EIA are aware that plants do not have an incentive to comply until after a violation has been discovered. They generally respond first by pointing out that the statute requires that they seek voluntary compliance and that they have no choice. However, the statute does not lay down any guidelines defining voluntary compliance. There have been no court challenges of the department's interpretation of it, suggesting that the requirement is not a legal constraint on EIA behavior. More likely it is a political constraint or an EIA preference for voluntary compliance. Some EIA personnel did, in fact, voice a preference for voluntary compliance, arguing that while it may sacrifice something by failing to provide incentives until after a violation is discovered, it more than compensates by the rapport the EIA has with the sources in the state. They fear that if the EIA were to adopt a punitive rather than a corrective approach to violations, industry would cease cooperating with the state. Problems such as having to get a court order to conduct source surveillance, or finding a lack of candor among plants regarding their emissions were mentioned as possible results of a departure from the voluntary compliance approach.

Are these worries justified? Again, one must look beyond New Mexico and compare its program with those in states which do not practice voluntary compliance. Based on a small informal sample, it is true that the EIA is generally well regarded by the operators of air pollution sources

in New Mexico. In particular, if among them a popularity contest between
the EIA and the EPA were held, there is little doubt who would lose. Nonetheless, it is conceivable for an air pollution agency to be unpopular
without being crippled by noncooperation.

Defining and Determining Compliance

As noted above, establishing a definition of compliance is the first
essential task of enforcement. Compliance is a word that is often care-
lessly used, but making the concept more specific raises questions that
are perhaps not immediately obvious.

Most New Mexico regulations are written in terms of a mass emission
rate per unit of input, such as pounds per million Btu of fuel, or per unit
time, such as pounds per hour. However, it is a rare source that achieves
a constant rate of emissions; even under normal conditions (no equipment
failure) emissions can vary considerably within hours. With rare exceptions New Mexico's regulations, like many Federal regulations and those
of many states, take no account of such variability during operation, and
by implication leave it to the EIA to give "compliance" an operational
definition.

In a world of perfect information one would have to confront this
ambiguity head-on. Specifically, for how long and by how much does a
source have to exceed an emission standard to be considered out of compliance? Is noncompliance a matter of degree, and should violations be
dealt with in proportion to their severity? Should the severity of a
violation be measured as the total mass of emissions above that allowed

by the regulation, and if not how should severity be measured? Under
what circumstances, if any, should sources be given credit for an emission
rate below the regulation, and how?

In practice, questions of this sort rarely come up, for two reasons.
In the first place, continuous information on emissions rarely exists.
Compliance has to be defined by what information is available, and this
implies that the definition of compliance for a source depends on how its
emissions are determined. For example, for most categories of sources
the most accurate information about emissions is acquired during a source
test, in which stack emissions are determined directly. A sample of flue
gas is taken out of the stack and the concentration of pollutant in the
sample is calculated. This is combined with estimates of total stack gas
flow to arrive at a total mass emission rate for the pollutant, which can
then be expressed in whatever way is convenient. Thus a source test is a
snapshot, or more accurately, three snapshots of emissions taken on the
same day. The average emission rate of the three runs is taken to be
result of the test, and the determination of compliance is a simple matter
of comparing this number with the number in the regulation. This procedure provides an estimate of the short-term average emission rate, which
is evidently the criterion ordinarily used to determine whether a violation
has occurred.

However, conducting a source test is an expensive proposition. The
amount of effort varies, of course, depending on the complexity of the
source, but can be as much as six man-weeks (the amount of effort required
for a source test is discussed in more detail in the next section). To be
able to monitor the performance of more sources, the department uses

engineering inspections. For some, but not all, sources such an inspec-

tion can give an indication of whether a source is out of compliance. If

it appears to be, a source test or some corrective measure can be ordered.

The use of inspections, while probably a more cost-effective use of the

EIA's time, provides less information than source tests.

While a source test does give an indication of the degree of a viola-

tion, this information is not factored into the definition of compliance.

In fact it cannot be, owing to the reliance on voluntary compliance.

Because a plant must be given a chance to come back into compliance volun-

tarily, it cannot be penalized according to the magnitude of a violation.

In this approach the severity of air pollution violations is measured in

entirely different dimensions, principally the duration of the period of

noncompliance and the apparent willingness of the source to cooperate.

In addition to surveillance, the EIA relies extensively on self-

reporting, especially for certain source categories and certain pollu-

tants.[2] Table 5-1 shows the usual method of determining compliance for

each industrial source category and pollutant.

As shown, the Agency depends on self-reporting to determine compli-

ance with the sulfur emission regulations for the petroleum refineries,

natural gas processing plants, acid plants, and smelters in the state.

At all these plants, the sources rely on a sulfur balance to estimate

---

[2]In the monitoring of initial rather than continuous compliance,
the Agency relies more extensively on self-reporting. For those regulations with effective dates, the regulations required schedules of
compliance, and the Agency required sources to submit progress reports.
As noted, variances and assurances usually were also conditioned on the
submission of periodic progress reports.

Table 5-1. Methods of Determining Compliance with Air Quality Regulations in New Mexico

Category	Pollutant	Method	Self Reporting?	Number of Source Test in Category between 1/1/76 and 7/31/78
Natural gas processing	Sulfur	Materials balance	Yes	2
Petroleum refining	Sulfur	Materials balance	Yes	2
Petroleum refining	HC	Inspection	No	NA[b]
Utility Steam Generators				
Coal-fired	Sulfur	Source test	No	1
Coal-fired	NO$_x$	Source test	No	0
Coal-fired	Particulate	Source test	No [a]	6
Gas-fired	NO$_x$	Source test	No	6
Oil-fired	Sulfur	Source test	No	2
Oil-fired	NO$_x$	Source test	No	2
Oil-fired	Particulate	Source test	No	2
Asphalt processing	Stack emissions	Source test	No	6
Asphalt processing	Fugitive emissions	Inspection	No	NA[b]
Woodwaste burning	Smoke	Opacity reading	No	NA[b]
Sulfuric Acid plants	Sulfur	Materials balance	Yes	2
Nonferrous smelters	Sulfur	Materials balance	Yes	0
Nonmetallic minerals	Stack emissions	Source test	No	12
Nonmetallic minerals	Fugitive emissions	Inspection	No	NA[b]

[a] The Four Corners Plant, which reports to the EIA on its particulate emissions, is an exception . See Chapter 6.

[b] Not Applicable.

emissions, which are assumed to be the residual required to make the sulfur content of plant outputs balance the sulfur content of plant inputs. The sulfur balance is a good example of a measurement method placing a limitation on the definition of compliance. The regulations for these industries are unusual in their specification of a time period over which compliance is to be determined: for natural gas plants, the emission limitation is not to be violated "at any time," while for petroleum refineries and sulfuric acid plants a 24-hour period is specified. However, a sulfur balance is not appropriate for measuring instantaneous emissions or even emissions over a 24-hour period. Because sulfur emissions are a small fraction (usually around ten percent) of the total sulfur processed in the plant, a small measurement error in any stream has a magnified impact, on a percentage basis, on the residual. This is true even if, as is the case for these plants, sulfur outputs and inputs are measured daily. Over time these errors average out and one can get a reliable reading from a sulfur balance, but on a daily basis a sulfur balance simply cannot give acceptable results. As it happens, the EIA requires sulfur reports from affected sources in these categories on a monthly or a quarterly basis, and what is reported is average daily emissions for the quarter. Thus, no matter what the regulation says, what is regulated is a 30-day or 90-day average.

For the other surveillance methods listed in Table 5-1, the EIA relies on its own personnel. Besides source tests, which have been described earlier, the Agency takes opacity readings for woodwaste burners and other sources of smoke, and conducts engineering inspections. Of all the surveillance activities the Agency relies upon, the opacity reading is

the least time-consuming and requires the least training. Such readings
are made by trained "smoke readers," usually volunteers or part-time
workers working out of one of the eight regional offices of the EIA. They
are called "environmentalists," and take a short course conducted by the
Agency to learn how to read opacity. It takes about half a day to perform
an opacity reading, with a test period of perhaps four hours. Typically,
a reading is taken every quarter hour, and the fifteen-odd readings are
averaged to obtain a test result.

An inspection is a somewhat more complicated affair, though much less
involved than a source test. As the name implies, the plant's operation
is inspected by an EIA engineer. He examines the visual quality of the
stack emission and operation of the control equipment, looking for problems such as torn or blinded filter bags, broken precipitator wires, or
insufficient water flow to a scrubber, and notes any fugitive emissions.
Strictly speaking an inspection cannot be used to determine compliance
except for fugitive dust. The Environmental Improvement Board has, in
effect, outlawed fugitive emissions, for it has required that all emissions
for asphalt processors and nonmetallic mineral processors exit through the
stack. This makes compliance with fugitive dust regulations easy to
determine during an inspection. The uses of inspections are far from
limited to a determination of fugitive dust, however. For particulate
emissions, an observer can often tell whether a violation is occurring.
The emission regulations for asphalt and nonmetallic mineral processors
are set sufficiently low that a visible emission probably represents a
violation. Thus, if the inspector finds a visible emission, either he can
order a source test to document the violation, or he can order the source

to conduct a source test, or, less formally, he can inform the source

that it is probably in violation and request that it take whatever action

may be required to eliminate the visible emission.

At the outset of this study, one of the findings anticipated was the

widespread use of inspections as a screening device for source tests, as

mentioned above. While this happens, it is surprisingly uncommon. Many

of the source tests are performed for air contaminants, in particular SO_2

and NO_x, for which determination of compliance by inspection is not easy

because the emissions cannot be seen. Since the beginning of 1976, 15 of

40 source tests measured emissions of pollutants other than particulate.

The use of screening inspections seems most appropriate for asphalt pro-

cessors and nonmetallic mineral processors, but even in these categories

one finds that between January 1, 1976 and July 31, 1978 only 10 percent

of the source tests was preceded within three months by an inspection

the result of which was a suspected violation. This casts a dubious

light on the assertion that source tests were made only when violations

had previously been uncovered by an inspection.

Allocation of Enforcement Activities

Performance of the tasks of enforcement is principally the function

of the Enforcement Section within the Air Quality Division of the EIA.

Every professional in the Enforcement Section has a technical background,

and the Section's main functions are to conduct source surveillance and

negotiate with plants about achieving compliance. Like other states, New

Mexico has had some problems with high turnover in its air quality program.

To the universal problem of low state government salaries, the duties of

enforcement add some unique conditions which cause turnover to be high.

The job involves a considerable amount of travel and requires one to spend a lot of time in unpleasant or possibly hazardous surroundings monitoring emissions. New Mexico does offer one advantage that many other states cannot, for many people will gladly trade salary and other benefits to be able to live in Santa Fe. Despite this, it is estimated that the average employment duration of the staff of the Enforcement Section is about 18 months. Inasmuch as it takes about six to eight months to become proficient at the tasks of surveillance, each staff member can be expected to produce about a year of useful work before departing. The Enforcement Section consists of seven professionals, all with technical backgrounds, and for six surveillance is their principal activity. These six can be broken further into a source testing team of three members (a team leader plus two technicians) and three inspectors. The seventh member is the section chief, whose chief duties are the supervision of the Section plus communication and negotiation with sources when a violation is discovered. In addition, the Air Quality Division requires approximately 80 percent of the time of one lawyer from the EIA Legal Section, mostly for enforcement-related matters.

Several years ago the Enforcement Section put considerably more emphasis on source testing. Between early 1971 and mid-1974 the Section gradually built up its testing capability; until early 1973 there was one source testing team, then until late 1974 there were two teams, and finally, during 1975 and 1976, three teams were used. The reason for the buildup was a desire to test every major source in the state at least once. By late 1976, this goal had been reached, and thereafter it was felt that enforcement resources could be more effectively and efficiently

spent conducting inspections. Besides, with three teams there were
difficulties in scheduling the use of the available equipment, so that
three teams were not producing three times the number of tests that one
team had produced earlier. For these reasons it was decided to retreat
from a three-team to a one-team operation, and the Enforcement Section
has operated at this level since the beginning of 1977.

Nonetheless, even now about 75 percent of the resources of the
Enforcement Section are devoted to surveillance, while the remainder are
devoted to the correction of violations. To be sure, this breakdown may
be a bit misleading because of the dual function of surveillance. While
source surveillance is necessary to find out what sources are doing on a
routine basis, it is also necessary in order to find out whether previously discovered violations have been corrected. It has not proved
feasible to classify individual surveillance events into these two categories, although it appears evident that the surveillance agenda is
sensitive to the compliance status of the sources in the state. This is
discussed in more detail momentarily, but first we turn to the cost of
the tests.

Costs of Surveillance. In this subsection are given estimated administrative costs, in 1978 prices, of source tests and inspections. As
shown in Table 5-2, the fixed cost of maintaining a three-person source
testing team is estimated to be $4,800 per month: $3,000 for salaries,
$1,200 for benefits and overhead, plus $600 for equipment. The equipment
charge reflects the use of $35,000 worth of equipment with approximately
a five-year life span. To convert this into a cost per test, we need the
output of a source testing team in tests per month. Depending on the

Table 5-2.   Unit Administrative Costs of Source Tests and Inspections

(1978 prices)

(a)   Source Tests

Monthly Fixed Costs – One Source Testing Team (3 people):

Salaries, benefits and overhead	$4,200	
Equipment	600	($35,000 in equipment amortized over 5 years.)
Total	$4,800	

Variable Costs per test:

Per diem	$ 300	(3 days at $32/person/day)
Vehicle	50	(300 miles at 17¢/mi.)
	$ 350	

Average Cost per test:

$2,750 at 2 tests per month

$3,900 at 1.2 tests per month

(b)   Inspections

Monthly Fixed Costs:

Salary, benefits and overhead	$1,680
Equipment	--
Total	$1,680

Variable Costs per Inspection:

Per diem	$ 64
Vehicle	50
Total	$ 114

Average cost per inspection: $530 (at 4 tests per month).

complexity of the source and its distance from Santa Fe, a source test will require as much as two days of travel and three days of actual testing. (The extreme parts of New Mexico are a day's drive from Santa Fe.) In addition to the time in the field, there is also a significant amount of office work associated--performing the calculations and writing up the test results (a typical source test report runs thirty to fifty pages). The Agency objective is two tests per month for each source testing team. However, since the Agency cut back to one-team operation at the beginning of 1977, 21 source tests were conducted in 17 months (through May, 1978) for an average of 1.2 source tests per month.

If the performance of the source tests is the sole function of the source testing team, then the fixed cost allocated among source tests comes to $3,900 per test. Added to this is an average variable cost of three days for testing and 300 miles of travel, for a total estimated source test cost of $4,250. On the other hand, if the EIA objective of two tests per month is in fact met and the source testing team members spend part of their time engaged in other, unspecified tasks, then the average cost of a source test is $2,750. We take these estimates to bound the cost of the average source test. By comparison, one source which purchased a source test from an engineering firm reported a price of $3,800.

Inspections, of course, are considerably less expensive. Making similar assumptions as above we find a cost of $1,680 per month to support one engineer-inspector. The EIA standard is four inspections per month per month per inspector, which, with travel and per diem, yields an estimated cost of $500 to $550 per inspection.

The Surveillance Schedule. How does the EIA determine which source

to visit next? Although there appears to be only one formal requirement,

the Agency also uses several rules of thumb to guide the selection of

sources. The requirement is concerned with frequency: above all the

Agency attempts to visit every source (except those which practice self-

monitoring) at least once per year. This requirement is imposed by EPA,

which considers annual surveillance to be acceptable for SIP implementa-

tion. On the Compliance Data System (CDS), EPA's computer system used

to monitor the compliance status of major stationary sources, any source

for which the status has not been updated within the past year is put in

the category of "unknown compliance status." Of all the sources in New

Mexico outside of Bernalillo County, approximately 16 percent were listed

in the CDS Report of May 2, 1978 as having unknown compliance status.

The Agency also imposes guidelines on itself for inspection frequency

for certain source categories. For woodwaste burners, the Agency attempts

to perform opacity readings at least every other month. A sample of six

woodwaste burners showed an average of 2.1 months between opacity read-

ings. In addition, the Agency attempts to inspect every portable asphalt

plant at every temporary site. Typically, these plants are in operation

at a particular site for a period of from several weeks to several months.

However, it is difficult to obtain an estimate of the EIA's success in

visiting all sites, because the material on file reports primarily on the

sites the Agency knows about and has inspected. Although all asphalt

plants in the state are supposed to report to the EIA before they set up

temporary facilities, this is not always done.

Surveillance activities are also directed by the receipt of complaints. The EIA does not appear to receive many complaints about
industrial stationary source emissions, but all must be investigated,
and often the complaining party is informed of the outcome of the investigation. Historically, complaints have arrived at EIA headquarters at the
rate of about 1.5 to 2 per month, and have been concerned mostly with
asphalt processors and woodwaste burners, as shown in Table 5-3. Inter-
estingly, the complaints in each category are directed primarily at one
producer within the category. Moreover, the frequency of complaints about
these plants seems to be due more to the location than the performance of
the plant with respect to emissions (this point is pursued in the next
chapter).

Another consideration guiding the schedule of surveillance is travel.
If an inspector must make an overnight trip to, say, Hobbs, an attempt
will be made to see as many sources as possible in the vicinity of Hobbs
or on the way to and from there.

Once the schedule of plants to be visited has been determined,
visits are made in the daylight hours, and the plants are usually notified
a week or two in advance, and always advance notice is given when a source
test is scheduled. The reason plants are notified is to ensure that they
are operating, so that Agency personnel may avoid wasted trips. About
the only time the Agency conducts a surprise inspection is if there is
reason to believe that a violation is occurring and the plant may attempt
to conceal that fact. It is often argued that if a plant is notified in
advance of source surveillance, then it has a chance to cheat, or make
temporary changes to the operation of its equipment in order to appear to

Table 5-3. Complaints by Source Category, 1971-1977

Category	Number of Complaints
Petroleum refining and natural gas processing	5
Utility steam generators	7
Asphalt processing	45
Woodwaste burners	32
Sulfuric acid plants	5
Nonferrous smelters	7
Nonmetallic minerals	10

be in compliance. Where plants are fined for a lack of compliance there will surely be two strong incentives to do just that: the fine itself, plus perhaps the extra cost of operating the equipment in such a way as to be in compliance. In New Mexico a plant need not worry about a fine, although it still may have to incur operating expenses which could have been avoided had not a violation been discovered. Hence, the incentive for cheating may be relatively small, so that surveillance in New Mexico may give an unbiased picture of actual emissions.

The Imposition of Sanctions: A Comparison

Since 1971, the EIA has instituted court proceedings against five plants. Three cases were settled out of court and dismissed, and two actually went to court. Of the plants involved, four were woodwaste burners and the fifth was the Town of Taos, which was involved with the state in a dispute over solid waste incineration. In this section, case histories for the two cases which went to court will be outlined, and compared with two other case histories of mineral processing firms. These cases will give a flavor for what voluntary compliance means in practice.

Jackson Sawmill. Jackson was a small family-owned sawmill located in the Rio Grande Valley near Espanola, which has since gone out of business. On October 30, 1970 the company received an assurance of discontinuance to upgrade the performance of its burner. The assurance expired on December 31. In the next three months, the burner was monitored on seven occasions, and six times was found in violation of the opacity regulation (at that time 40 percent). The average of the seven opacity readings was 79 percent. On March 25, 1971 a complaint was filed, stipulating that Jackson had been in violation since the beginning of the year. This complaint was settled out of court, and a consent order was issued on June 28 dropping the charges. Certain modifications were made to the control equipment in the interim; in fact, from mid-April until August the EIA recorded six opacity readings at the mill, with no violations. Average opacity was 17 percent.

Apparently no more readings were made between August and the following February, but a series of visits between February and May, 1972 disclosed that the plant was again in violation (three out of four readings showed violations; average reading, 65 percent). On May 11 the Company was sent a letter notifying it of four violations since the beginning of the year. A week later Jackson replied, claiming that it had spent $40,000 for pollution control in the last two years, and that technology did not seem to be available to reduce the opacity at the mill.

On September 28, a second complaint was filed. That fall and winter the Agency stepped up the frequency of opacity readings on Jackson's burner, apparently to bolster the State's case. Nine readings were made between September 1 and March 1, 1973 with six of those occurring in February. The trial was held on March 6 and 16, 1973. In June the parties entered into a consent order which stipulated that if the violations continued after November 1, a $5,000 fine would be levied. In September and October two readings of 5 percent were recorded, but by December the burner was again out of compliance. Shortly after this the mill closed down. Apparently the reason was a soft market, not the EIA.

Otero Mills. Located in Alamogordo, Otero is a mill with two wood-waste burners, though the smaller is used only on an exceptional basis. During 1970, the mill was visited on three occasions, with one violation on each burner being reported. After the last visit, a warning letter was sent to Otero asking the company for better compliance with the regulation. During the first four months of 1971, three more inspections were made, with one violation found on the large burner.

In early May something happened to the large burner, causing it to be out of compliance consistently. Violations were reported on May 3, May 6, May 7, May 13, May 18, May 19, and May 24 (the reason that the EIA could make so many readings on the mill was that there is a regional EIA office in Alamogordo). In addition, when the small burner was used it seemed to be in violation also. Readings of Ringelmann 2.5 (50 percent opacity) and 4.9 (98 percent opacity) were recorded on the small burner on July 7 and September 24.

On November 22 the EIA received a letter from New Mexico Citizens
for Clean Air and Water complaining of frequent violations on both burners.
An inspection the following day found the small burner in violation and
the large burner in compliance. Nonetheless, the Agency requested, under
the auspices of voluntary compliance, that by Christmas the firm cease
using the small burner altogether and increase the overfire air on the
large one. Otero agreed, and on December 13, 1971 the Agency was informed
that the firm had ordered another blower for the large burner, and that
the small burner was being shut down and dismantled. At about that time
the manager of a local campground complained that debris from a local
sawmill was falling out on his premises, damaging his business. An EIA
investigation determined that Otero was the cause.

Eleven routine smoke readings during 1972 yielded seven violations.
Performance, in fact, seemed to deteriorate over the year, so that by
December and January, 1973 the results of a special series of readings
by the local environmentalist reported an average opacity of about 70
percent. A letter was sent to Otero on January 17 requesting that the
Company outline by the end of the month its plans for coming into compliance. The Company's reply claimed that the increase in overfire air
did not help, and that an additional blower had been ordered.

In early February, 1973 the EIA Legal Section sent Otero a letter
noting ten violations since September and asking the company to submit a
satisfactory plan for eliminating the violations or face legal action.
The company responded verbally, requesting a meeting in April to discuss
Otero's problems. On April 11 a memo to the EIA headquarters from the
local environmentalist reported six violations in nine readings made

between February 1 and April 9 (Average opacity: 47 percent). By this

time the second blower had been installed, but the mill was still fre-

quently in violation. On the next day the owner of the mill met with EIA

officials in Santa Fe. At this meeting he described some changes which

would improve the performance of the burner, but he refused to sign an

assurance of discontinuance.

The following August, the company decided to apply for a variance

and requested application forms from the Agency. The completed petition

was received in mid-October. Although 30 of 44 observations made in the

past two years had shown violations, the EIA was initially favorably dis-

posed to this request, and the Board gave tentative approval and scheduled

a hearing for December 1. Unfortunately for Otero, however, an EIA

inspector visited the mill on November 14. The burner was not in opera-

tion that day, but an inspection of the facilities revealed evidence of

poor operating practices and poor maintenance of equipment, giving the

Agency reason to reconsider its initial approval of the variance request.

During the public comment period the Board received numerous letters from

the citizens of Alamogordo regarding the proposed variance, mostly un-

favorable. This variance was denied, and the reasons given for the denial

were the sloppy working conditions, Otero's apparent lack of willingness

to improve, and the apparent lack of economic harm which would follow the

denial.

Meanwhile, performance of the burner did not improve. Five readings

made between December and early March, 1974 all resulted in violations,

with an average opacity of 53 percent. A certified letter was sent noting

these five violations, asking the company to reply. Otero's response was to predict that compliance was right around the corner, attributing the violations to typical problems encountered with recently installed equipment. However, five more readings extending into May showed only marginal improvement: an average opacity of 47 percent.

At last, on May 22 a complaint was filed, citing 36 separate violations between April 26, 1972 and April 22, 1974. The trial was set for September originally, but was postponed several times, finally taking place on May 24, 1975 (Otero's lawyer was a state senator, and he requested a number of postponements because of the press of senate business). In the meantime, the burner was shut down in February and the mill began to stockpile its waste. Because the burning had ceased the complaint was dismissed, but with a warning any further violations by the burner would lead to fines.

Three years later, on June 20, 1978, the mill shut down for good. Otero had been getting timber from the Mescalero Apache reservation, but the Indians refused to renew the lease. One month after closing, Otero's immense woodwaste pile ignited spontaneously, and the fire was not put out until mid August. However, most of the waste remains, a fire hazard and a potential smoke source.

Grefco, Inc. Grefco has a perlite plant located at No Agua, in the north central part of New Mexico near the Colorado border. The Board promulgated a regulation for perlite processors on September 1, 1971, and on January 10, 1972 the company requested an assurance of discontinuance for twelve months to obtain and install a baghouse filter. Soon thereafter, the company began to experience delivery difficulties, and the EIA agreed to postpone the date of compliance from January 10, 1973 until

June 30, 1973. The following February the company reported that the
delivery date had been moved back again, until May 1973. No compliance
date was estimated, but it was obvious that the June 30 deadline was no
longer feasible. In May, delivery was postponed again until July 30. At
this time the company estimated a compliance date of October 31. On
September 23 the equipment arrived at last, with compliance then estimated
for January 15, 1974, one year late.

However, it turned out that the equipment that Grefco was installing
was inadequate to control the emissions at the plant. Perlite processing
results in two gas streams from which particulate must be removed--a "hot"
stream and a "cold" stream. At the Grefco plant, these streams were combined to be treated in a single baghouse. This approach is considerably
cheaper than keeping the gas flows segregated. But when a cold air stream
is brought into contact with a hot air stream, the change in temperature
can cause a condensation of much of the water vapor contained in the hot
air stream. If the combined stream, now laden with moisture, is introduced
into a baghouse, the moisture collects on the surface of the bags and clogs
them up. The resulting "bag blinding" prevents any air from penetrating
a bag, rendering it useless as a filter. Instead of keeping the gas
streams separate and using two baghouses, Grefco attempted to get by with
a quick fix, principally better insulation of the duct work to prevent a
greater temperature drop than absolutely necessary.

Bag blinding problems appeared immediately after the baghouse came
on line in early 1974, but the company promised to have them solved by
July 1, 1974. And in fact, late in the year a source test revealed that
the stack was in compliance for the first time, although considerable

fugitive dust emissions were reported. An inspection on May 23, 1975

found that fugitive dust controls had not been installed.

Moreover, at about this time, it was discovered through complaints

that Grefco was operating the baghouse only about half the time. Each

day, the baghouse would be operated until the bags became blinded, where-

upon the pollution control system was bypassed altogether. At the time

the plant was operating on a two-shift basis, and the blinding occurred

after about eight hours. During the second shift the bags would be

cleaned in preparation for one-shift operation the following day. Upon

being notified that this was an unacceptable practice, the company on

July 22 filed an "801" (upset) request for permission to operate the

plant in the same fashion until the annual shutdown in mid-August. The

EIA complied, but shortly after the plant was restarted, Grefco requested

an extension of the Agency's indulgence until December 1.

On November 7 the EIA acceded to this request, but informed Grefco

that this was its last chance. Citing a lack of voluntary compliance,

the Agency threatened legal action if compliance was not achieved by

December 1. The deadline passed without compliance having been achieved,

but there was some evidence of progress. On that basis the company was

given until January 31, 1976 to cease its violations, and on January 22,

1976 a letter from the EIA informed the company that violations after the

end of the month would no longer be exempted by an upset report. The

company agreed, saying that it would immediately cut back to a one-shift

operation.

One interesting and relevant question which comes up at this point

is whether Grefco retreated to one-shift operation solely at the behest

of the EIA. It is not likely that it did. The company had just opened
a new plant elsewhere, and probably did not need two-shift operation at
No Agua at that time. If two-shift operation had been otherwise profitable, it is almost inconceivable that the company would so readily have
agreed to the EIA's request. In any event, an inspection on April 1 disclosed that the plant did indeed shut down after one shift. To be sure,
the filter bags still blinded daily, but the plant was in full compliance.

Compliance was only temporary, however. In a source test in July
the stack emissions were found to be 75 lbs per hour, compared to an
allowable emission of 33 lbs per hour, with substantial fugitive dust
emissions. Upon notification, Grefco attributed the excess emissions to
bag failures due to abrasion, and promised better maintenance. On
September 23 the Agency informed Grefco that this was not a satisfactory
response. The Company was ordered to correct the fugitive dust problem
and arrange for a source test by mid-November to demonstrate compliance.
In the ensuing exchange of letters, Grefco and the EIA sharply disagreed
over whether the Agency had the authority to order a stack test, but at
a meeting on December 2 the company agreed to conduct a test "voluntarily."
Nonetheless, the EIA did conduct a source test on March 21, 1977, which
yielded emissions of 6 lbs per hour, well within the regulation.

Since that time the plant has been inspected three times, most
recently on June 29, 1978, and, except for one minor fugitive dust
violation, has been found to be in full compliance, even though two-shift
operation has resumed. To prevent bag blinding for two shifts has required
a relatively expensive maintenance program. Blinding is delayed by more
vigorous bag shaking, but this procedure considerably shortens bag life.

In addition to the cost of the bags, one half to three quarters of an hour per shift is required to inspect and replace broken bags. This represents about an 8 percent loss of output.

J. H. Rhodes and Company. Rhodes operates a pumice processing plant on the outskirts of Santa Fe, about two miles from EIA headquarters. Pumice plants are governed by the same regulation governing perlite plants, which, for a plant with the capacity of the Rhodes plant, calls for emissions of 10 lbs per hour or less. Probably because of its nearness to Santa Fe, the plant already had a baghouse filter system in operation when the regulation was promulgated (September 1, 1971).

The first formal contact between the EIA and Rhodes was a source test conducted on March 1, 1972, the result of which was an estimated emission rate of 23 lbs per hour. Upon notification of a violation, the Company on April 10 informed the EIA that new bags were to be installed by the end of August. However, there was a delivery delay of several weeks, and as a result the bags were not replaced until close to the end of the year. On January 8, 1973 the company conducted a stack test and reported emissions of 3.3 lbs per hour from the stack, well within compliance.

Meanwhile, the company had been notified on several occasions of fugitive dust violations, and on December 15, 1972 Rhodes requested an assurance of discontinuance until May 31, 1973 to correct the fugitive dust problem. The EIA agreed. On May 18, however, Rhodes requested an extension of the assurance until the following May, but the Agency refused.

On June 12 a source test found emissions of 12 lbs per hour, plus significant fugitive emissions, and notified the company of violations of both the stack and fugitive dust regulations. The company responded that new bags had been ordered (they were ordered on June 12, the day of the stack test). After the bags were installed the plant was apparently in compliance, but no formal inspection was made until the following April, when compliance on both stack and fugitive emissions was observed.

In response to a complaint, an inspection on January 15, 1975 uncovered a fugitive dust violation and a probable violation of the stack regulation. The company's response blamed it on a small group of broken bags that needed replacing, and the Agency told Rhodes in the future to file an upset report under such circumstances. The complaint was from a local homeowner who lived in a subdivision near the plant. As more and more houses in the subdivision became occupied, the company found that it could no longer operate the plant in violation without risking complaints by the homeowners. After a pair of complaints in May the Agency on June 11 informed Rhodes that it was not correcting malfunctions quickly enough. Previously, the company had been changing bags only on Saturdays when the plant was down, so that on average a bag malfunction caused excess emissions for half a week. The EIA informed the company that if bag malfunctions were not repaired or replaced the day of occurrence then it would be considered a violation of the regulation.

Inspections following three more complaints in the next year disclosed violations, the first two of which were rapidly corrected. After the third complaint, a source test was given in June 1976, which reported emissions of 44 lbs per hour. On July 9 the violation was reported to the firm, with a promise that further violations would lead to court action. A week later a meeting promised compliance by late September.

On January 17, 1977 a source test by a consultant to the company indicated emissions of 8 lbs per hour, suggesting that the plant was back in compliance. However, at an inspection by the EIA on March 4 the baghouse was found to be puffing badly, although the plant was found to be in complete compliance with the fugitive dust regulation; in fact, it was spotless. Agency officials suspected that the plant was cleaned up for the benefit of the company president, who had flown in from Chicago specifically to be on hand for the inspection, and indeed an inspection three weeks later found a probable stack violation, plus four major fugitive dust leaks.

Because of the puffing baghouse the Agency conducted a source test in April, which revealed emissions of 20 lbs per hour. The case was shortly afterward referred to the Legal Section, which on June 15 informed Rhodes of the likelihood of court action unless the violations ceased immediately. Five days later Rhodes replied, claiming that the company had been unfairly singled out, and that the Agency's test result was incompatible with the company-conducted test in January. It disclosed that a consultant had been retained to conduct another source test within a month to demonstrate compliance. Six weeks passed and the EIA had not received word from the company on the outcome of the test. When the Agency inquired, the company admitted a test result of 17 lbs per hour.

Shortly after their test, the company apparently solved the problem it had been experiencing with the control equipment. A source test on September 20 and an inspection on September 23 showed full compliance, including estimated emissions of 1.5 lbs per hour. The plant has apparently been in compliance ever since.

The compliance record for these four cases is summarized in Table 5-4. It is clear from the table that a considerable amount of noncompliance will be endured before legal action is initiated. It also seems clear that the violations by the two sawmills was no more serious than the pumice and perlite processes, and yet legal action was pursued only against the sawmills. Why is this?

Table 5-4. Compliance and Legal Action at Four Plants

(times in months)

	A	B	C	D
Jackson Sawmill	36	3 (first) 21 (second)	a/	a/
Otero Mills	96	47	56	70%
Grefco, Inc.	78	none	62	65%
Rhodes, Inc.	78	none	66	b/

A - Length of record

B - Elapsed time from first discovery of violation to the initiation of court action.

C - Elapsed time from beginning of record to the point compliance was consistently achieved.

D - Estimated percent of interval C that the plant was not in compliance

---

^aPlant shut down before initial compliance was achieved.

^bFor about 14 months of the interval the plant was probably in compliance. For about 13 months it was not. For the rest of the time compliance was intermittent.

One possible explanation is the amount of evidence. Because the woodwaste burner regulation is written in terms of opacity, a violation can be determined by an opacity reading, which is quite simple and inexpensive to carry out. The EIA performed 46 opacity readings on the Jackson mill between 1970 and 1973, finding 27 violations. Even more readings were taken at Otero: almost sixty readings, two thirds of which were violations. Thus the Agency had amply documented the fact that both mills were usually in violation. While a similar conclusion seems very plausible for the two mineral processors, hard evidence was much more difficult to come by. The perlite and pumice regulation is written in terms of mass emission rates; therefore, evidence of a violation can only be obtained through a source test. Only a small number of violations could definitely be attributed to each one of these firms. Generally speaking, it is probably no accident that every complaint filed by the EIA has concerned an opacity regulation.

Jackson and Otero could also have run afoul of the EIA because of the nature of the voluntary compliance: these companies simply did not have anything else to offer. According to this view, either they could not or would not take the steps necessary to achieve compliance, and without evidence of progress the Agency had no choice. While the two mineral processors were just as often in violation, they could always hold out the promise of compliance in the future: new filter bags installed, dust leaks plugged, insulation installed, and the like. And indeed, however long it did take, compliance has apparently been achieved at both plants.

Chapter 6

CONTINUOUS COMPLIANCE

In the last chapter the concern was with the manner in which the
state enforces its air quality regulations. In this chapter we turn our
attention to the sources themselves, to examine how they have responded
to the regulations on a continuous basis. Ideal information for this
task is the same as the information which would be desirable from the
point of view of enforcement itself: a profile of emissions over time
for each source. As pointed out in Chapter 5, however, such information
rarely, if ever, exists for any source.

The kind of information which does exist depends on the source
involved. For self-reporting sources the record gives 30 or 90-day
averages of sulfur emissions, based on a calculated materials balance.
For other sources we have a record of surveillance events consisting of
inspections, source tests, and opacity readings. From this record it is
desirable to be able to make inferences about the continuous performance
of the source, and it is worth considering for a moment what inferences
are possible. First, we note that except for woodwaste burners very
little information exists regarding the degree of compliance or violation
with the regulation because source tests are so infrequent. For asphalt
processors and nonmetallic mineral processors most of the information
comes from inspections, the result of which is either "this source appears
to be in compliance," or "this source appears to be in violation."

With this information about the best we can expect to do is estimate the
frequency with which a source will be in compliance.

But is even this a reasonable expectation? Perhaps not. For when
one asserts that the frequency of violation and compliance among the
surveillance record is an estimator of the overall-frequency of compliance
of the source, one assumes tacitly that (i) conditions have not changed
over time; and (ii) the surveillance record is a random sample of the
performance of the source. Neither of these assumptions can be made
with confidence. The two mineral processors discussed in Chapter 5 are
examples of sources whose current compliance status is probably much
better than is suggested by the historical record. As for the later
assumption, we have observed first that the Agency often suspects a
violation before an inspection and secondly that a source which knows
about an inspection or source test in advance may act differently than
it would otherwise. Both are reasons to expect bias, albeit with opposing
sign.

Nonetheless, let us suppose that these two assumptions (no change
over time and the randomness of the surveillance record) are met. We
still find that the estimates of the frequency of violation is for most
sources surprisingly imprecise, owing to the relatively small sample
sizes.

Each surveillance event provides a reading of whether a source is
in compliance. Assuming behavior is stochastic, the surveillance event
is a Bernoulli random variable (where 0 represents "compliance" and 1
represents "violation"). Suppose for a particular source we have a
record of n inspections with outcomes X_1, X_2, ..., X_n. An unbiased
estimator $\hat{p}$ for the frequency of violation p is about what one would

expect:

$$p = \frac{1}{n} \Sigma X_i = \frac{S}{n} ,$$

where S is the number of violations.

The standard error of this estimate is

$$s.e. (\hat{p}) = \left[\frac{\hat{p}(1 - \hat{p})}{n - 1} \right]^{1/2}.$$

For example, if in five inspections two violations are observed (by no means an unusual case), than $\hat{p}$ = 0.4, but a 95- percent confidence interval for p is (0.05, 0.85). In other words, whether the true frequency of violation was 5 percent of the time or 85 percent of the time, it would not be that unusual to find two violations in five inspections. This is a rather wide interval.

Source tests, opacity readings and materials balances estimate the extent of compliance or noncompliance, rather than just whether a violation has occurred. This additional information usually makes possible more precise estimates of the frequency, as well as the seriousness, of noncompliance.

Despite the problems which attend making inferences from compliance records, they do provide useful information on the behavior of plants and, probably, insight into the enforcement process. In this chapter this record is presented for a sample of sources in each category and some thoughts are offered on the factors which may affect compliance.

Records of Compliance for Stationary Sources

Self-reporting sources. As pointed out earlier, oil and gas plants, smelters and sulfuric acid plants estimate their sulfur emissions by materials balance and report to the EIA. There is no record of any one

of these plants ever having been cited for an air quality violation,

and only on a couple of occasions have these sources reported emissions

which were in excess of the regulation.

One of these occasions involved a so-called "declining emission"

gas plant. One provision of the regulation for gas plants exempts those

plants for which production is expected to decline such that before

1981 the plant would cease to be subject to the regulation (i.e., sulfur

throughput would fall below 7.5 tpd). To qualify, the operator submits

a schedule of expected sulfur emissions. The plant in question had sub-

mitted such a schedule, but a turnaround in the amount of gas processed

by the plant has postponed indefinitely the date when total sulfur release

would fall below 7.5 tpd. Accordingly, the plant now operates under

an assurance of discontinuance while its sulfur plant is upgraded to

come into compliance with the regulation, and the excess emissions reported

have been within the terms of the assurance.

All other gas plants in the state which are required to have sulfur

removal units appear to be in compliance, with average sulfur capture

efficiency of 92 percent. Likewise, the sulfuric acid plants and Phelps-

Dodge smelter report compliance with their regulations (95 and 90 percent

sulfur capture, respectively).

The Kennecott smelter has reported, on occasion, average quarterly

emissions in excess of the regulation, which calls for 60 percent control.

This regulation became effective at the beginning of 1975 and since that

time the collection efficiency of the sulfur unit has fallen below 60 per-

cent twice — 57 percent in the first quarter of 1976 and 44 percent in

the last quarter of 1976. On both occasions, the acid plant was inopera-
tive part of the time, and the plant was operating under an upset waiver.

Steam generating plants. As noted earlier, most of New Mexico's
utility generators are gas-fired. Until recently the EIA has paid little
attention to the gas or oil-fired generating plants in the state, but
recently the Agency has attempted to reach these sources and conduct
source tests. The gas plants, for which only NO_x is regulated, have
all tested at less than 0.3 lbs per 10^6 Btu input. The Agency has uncovered minor sulfur and particulate violations at an oil-fired plant
belonging to El Paso Electric.

Among steam plants, coal-fired plants account for the overwhelming
bulk of emissions of all kinds, and far and away the most important coal-
fired source is the Four Corners Plant. As discussed in Chapter 4, the
plant is not subject to any SO_2 regulation at present, and except for
Unit 1 operates under a variance with respect to NO_x emissions. The
particulate regulations call for emissions of no more than 0.135 lbs.
per 10^6 Btu input for Units 1, 2, and 3 and 0.5 lbs per 10^6 Btu input
for Units 4 and 5. The most recent source tests conducted by the Agency
on Units 1, 2 and 3 show that these Units are comfortably in compliance.
On Units 4 and 5, however, compliance can best be described as marginal.
Early in 1974 the Agency requested that Arizona Public Service perform
monthly source tests on Units 4 and 5 and submit these records to the
Agency on a semiannual basis. Pursuant to this request the company
conducted about 50 source tests on each unit between January 1974 and
March 1978. For Unit 4, 27 percent of these tests showed violations,
with an average emission rate for all tests of 0.49 lbs per 10^6 Btu input.

For Unit 5, 34 percent of all tests showed violations, with an average emission rate of 0.52 lbs per 10^6 Btu input. Table 6.1 suggests an apparent downward trend in the emissions from these two units, but it is not statistically significant.

Woodwaste burners. It was pointed out earlier that in order to achieve compliance a sawmill operator had to increase the temperature on his burner, or else find something else to do with his woodwaste. The usual way the former has been done is to increase the flow of air by installing overfire or underfire air blowers. An informal survey of mill owners indicated that each spent between $20,000 and $40,000 to install the necessary equipment.

Table 6.2 shows the average performance during the past five years for five randomly chosen woodwaste burners. Taken as a group these five burners were found to be in violation about a third of the time, but within the group the variation is considerable: individual mills are in violation from 13 to 55 percent of the time. Table 6.2 (b) indicates how performance (measured by average opacity) at these mills has changed over time. Before May 1, 1975 the regulation allowed a stack opacity of up to 40 percent. The period after May 1, 1975 we have broken into two subperiods, before and after January 1, 1977. As shown, all five burners were able to show considerable improvement when the regulation changed from 40 to 20 percent opacity. More recently, however, there has been a deterioration of performance at three of the five mills sampled.

Asphalt processors. Asphalt processors in New Mexico are subject to two regulations, one concerned with stack emissions of particulate and the other with fugitive dust. The fugitive dust regulation calls for

Table 6.1. Results of Source Tests for Particulate Emissions on
            Four Corners Units 4 and 5 by Year

(Emission Standard: 0.5 lbs particulates per 10^6 Btu input)

Unit 4

Year	Number of Tests	Number of Violations	Percent	Average Particulate Emissions (lbs/10^6Btu input)	Std. Error of Average
1974	10	4	40%	0.56	0.07
1975	7 a/	0	0	0.35	0.03
1976	18	4	22%	0.52	0.11
1977	10	4	40%	0.51	0.09
1978	4	1	25%	0.32	0.07

Unit 5

Year	Number of Tests	Number of Violations	Percent	Average Emissions (lbs/10^6 Btu)	Std. Error
1974	10	2	20%	0.44	0.04
1975	10	5	50%	0.73	0.12
1976	8	4	50%	0.62	0.12
1977	20	6	30%	0.42	0.03
1978	2	0	0	0.36	0.11

a/ Results of tests for the first half of 1975 were missing from the
    EIA file.

Table 6.2 (a). Average Performance of Five Woodwaste Burners 1974-1978

Mill #	Number of Readings	Number of Violations	Frequency of Violation	Average Opacity (percent)	Std. Error of Average Opacity
1[a/]	20	11	55%	42	3.5
2	13	6	46%	35	8.3
3	23	3	13%	15	3.0
4	15	6	40%	32	6.7
5	23	3	13%	15	2.8
Average			33%	28	

Table 6.2 (b). Trends in Woodwaste Burner Performance

Mill #	Average Opacity, Percent		
	Before 4/30/75	5/1/75 - 12/31/76	1/1/77-Present
Emission standard	40	20	20
1[a/]	45	36	44
2	34	24	55
3	21	12	18
4	53	23	20
5	24	10	9
AVERAGE	35	21	29

[a/] This mill executed an assurance of discontinuance in 1976 which allowed 40 percent opacity.

the elimination of fugitive dust; however, for portable asphalt plants
this is impossible, and the existence of a violation comes down to a
judgment regarding whether the dust is excessive. Whether the plant is
subject to a source test or an inspection, the result is a reading on
violations of both stack emissions and fugitive dust, although after
ian inspection one can only assert a probable violation of the stack
regulation.

The EIA's records for nine asphalt processing companies were examined,
and the results of inspections and source tests for those companies are
shown in Table 6.3. All the observations on one company were lumped
together; no attempt was made to distinguish among the individual processing plants within the company. In the aggregate, these plants were
found to be in violation with the regulations for about 40 percent of
the visits made by the EIA. This is slightly higher than the frequency
of violation among woodwaste burners, but what is striking about the data
for both categories is that one finds so much variation among individual
firms.

Nonmetallic mineral processors. Table 6.4 gives the performance
of six mineral processors in the state. However this information is
offered only for the sake of completeness; four of these plants have
been in situations in which the phrase "continuous compliance" is hardly
applicable. The first three are the plants discussed in the previous
chapter. The last entry refers to a plant which has been shut-down twice
and operated by three different companies since 1971.

Determinants of Compliance

The causes of variation in the performance of individual plants

Table 6.3. Performance of Asphalt Processing Firms 1972-1978

Firm	Number of Source Tests and Inspections	Stack Emission Violations		Fugitive Dust Violations[a]	
		Frequency	Std. Error	Frequency	Std. Error
1	14	0.29	0.13	0.21	0.11
2	3	0	–	0.67	0.33
3	6	0.67	0.21	0.83	0.17
4	7	0	–	0.29	0.18
5	5	0.60	0.24	0.20	0.20
6	10	0.60	0.16	0.50	0.17
7	5	1.00	–	0.20	0.20
8	7	0.43	0.20	0.29	0.18
9	7	0	–	0.29	0.18
Average violation frequency		0.39		0.36	

[a]The emission regulations for asphalt processing plants allow no fugitive emissions at all. However, this standard is too stringent to be taken literally, and inspectors at their discretion cite only "excessive" fugitive emissions.

Table 6.4. Performance of Nonmetallic Mineral Processing Plants

        1972–1978

Plant	Number of Source Tests and Inspections	Stack Emission Violations		Fugitive Dust Violations[a]	
		Frequency	Std. Error	Frequency	Std. Error
1	8	0.25	0.16	0.50	0.19
2	14	0.64	0.13	0.07	0.07
3	4	0.75	0.25	0.25	0.25
4	6	0.33	0.21	0	–
5	5	0	–	0	–
6	4	0.75	0.25	0	–

[a]The emission regulations for these plants also allow no fugitive emissions. As with asphalt processors, inspectors use their discretion and cite only emissions which appear excessive.

with respect to air quality regulations can be divided into two classes:
those factors which affect every plant in an industry and those specific
to individual plants. In this section we discuss these factors and their
relative importance.

Industry-wide factors. Though it is difficult to find factors which
are shared universally among the plants in an industry, some characteristics are held in common to a greater or lesser degree. Often all the
plants in a single industry are faced with the same regulation. As noted
above, the manner in which the regulation is written affects the way
compliance is determined, which may well affect the performance--or
apparent performance--of plants in the industry. The best example of
this is the performance of plants reporting sulfur balances, which have
always been in compliance. It is not that such plants are less than
honest because they are self-reporting; it is just that they are allowed
to average emissions over a 30 or 90-day period.

A second factor common to all firms in an industry is technology,
which is to say that the technical problem of reducing emissions is
roughly the same for all plants in an industrial category. For example,
not all gas plants employ exactly the same techniques for capturing the
sulfur released in the plant, but the process chemistry is roughly the
same and in any case very different from, say, woodwaste burning. All
woodwaste burners are faced with the same alternatives: increase the
temperature in the burner or find a user for the woodwaste. All asphalt
processing plants in New Mexico except two use venturi scrubbers (the
exceptions use baghouse filters).

One way technology appears to affect the pattern of compliance is in the breakdown between the cost of installing pollution control equip-

ment and the cost of operating it. When the capital cost is relatively

high one would expect a plant to attempt to postpone first-time compliance,

by seeking an amendment to the regulation, a variance or perhaps an

assurance of discontinuance. In New Mexico it is found that the companies

for which installation costs are high--petroleum refineries, natural

gas plants, copper smelters, the Four Corners Power plant--have accounted

for the bulk of variances and amendments. On the other hand, in those

instances where the installed cost of pollution control equipment is

low relative to the operating costs, one would expect that the principal

enforcement problem would be maintaining continuous compliance. That

certainly seems to be the case in New Mexico. As noted in the preceding

section, violations of the regulations are most often found among wood-

waste burners and asphalt processors, industries characterized by low

capital and high operating costs for pollution control. According to

one estimate provided by a mill operator, the cost of complying with

the woodwaste burner regulation is a total of $40,000 for capital costs

and $50,000 per year operating expenses. For asphalt processors, the

use of scrubbers means low capital and high operating cost. One very

rough estimate is that the air quality regulation added between 3 and

5 percent to the total costs of asphalt processors, but only about one

percent of that is the installed cost of equipment. More generally,

this difference may account for some of the variation in the way regula-

tions are written. For the high capital cost industries one finds

grandfather clauses for existing plants and future effective dates, both

of which tend to reduce the impact of initial compliance. Nothing com-

parable is provided in the woodwaste burner or asphalt processing requirements, thereby opening the door for negotiation. Even so, after

the source requirement has been established, the problems of maintaining

continuous compliance is the same in all states: Emissions must be moni-

tored and violations corrected. There is no reason to think that the

expense of source surveillance or its frequency will be notably different

in New Mexico.

Likewise, the supposed lack of a serious air quality problem in New

Mexico might be expected to contribute to a certain slackness in pursuing

regulatory objectives not found in other states. But this proposition is

dubious. For one thing, although ambient air quality in New Mexico is

generally better than national norms, it is incorrect to say that that

there is no cause for concern. The extraordinary visibility found in the

Southwest has deteriorated somewhat in the past few decades, a development

that has been linked to gradually increasing SO_2 emissions.[1] Besides, the

limited evidence available from other states suggests that EIA's affinity

for voluntary compliance and associated reluctance to seek legal remedies

for violations is far from unique. Although New Mexico is unusual in

having voluntary compliance required by statute, almost all states practice

it in fact.[2]

Even if New Mexico is perfectly representative, however, it is dangerous

to draw conclusions from only one observation. The "conclusions" of a case

---

[1]Trijonis, John and Kung Yuan, "Visibility in the Southwest: An Exploration of the Regional Data Base." U.S.E.P.A., Research Triangle Park, N.C., EPA 600/3-78-039.

[2]See "Profile of Nine State and Local Air Pollution Agencies," U.S. E.P.A., Office of Planning and Management, September 1980.

in the other. On the other hand, a perlite plant in Santa Fe has one of
the worst compliance records in the state. The woodwaste burner with
the best compliance record in the state is located in the middle of
Alamorgordo, although a burner in Espanola has been the subject of more
citizen complaints than any other source. Another example of the manner
in which location plays a role is the problems that asphalt processors
have in securing a water supply for their scrubbers in remote, arid areas.
The EIA has the authority to waive the regulation when water is unavail-
able, but to get a waiver the firm has the burden of demonstrating that
fact. Plants have been able to do this on few occasions. If water is
in short supply, the operator of the plant will quite naturally divert
the flow from his scrubber, a part of his operation which he probably
regards as unproductive anyway, to other uses.

Another set of attributes of the firm which may affect its compliance is its economic characteristics. This point was made in macro
when we considered the industry-wide factors above, but it works at the
micro level also, and for the same reasons.

A third factor contributing to compliance or noncompliance is the
most difficult to characterize: the attitudes of the owner or operator.
While staff members at the EIA--and air pollution officials generally--
appear to this writer to be a little too quick to ascribe differences
among sources in the degree of compliance and cooperation to attitudinal
differences, the fact remains that such differences are very real and
have very real impacts on the manner in which firms respond to regulations.
Of course, how one measures the influence of this factor is another
question entirely. There is a temptation to allow this factor to be

the residual; that is, all differences in responses which cannot be
accounted for by other causes is assumed to be the result of attitude.
This practice makes no allowances for an error term, and an independent
measure of attitudes is needed.

Judging from the evidence of Section 6.1, the variation among
plants and firms within the same industrial category is significant indeed.
Therefore, if one wishes to understand the response of a plant to an air
quality regulation, one cannot ignore factors specific to the individual
plant, and to the firm of which it is a part. This is unfortunate, because such information, much of which may be proprietary, is often difficult, if not impossible, to get.

Chapter 7

CONCLUSIONS AND POLICY IMPLICATIONS

This chapter contains some tentative conclusions about the current enforcement of air quality regulations for stationary sources, based on the experience of New Mexico. This will be followed by a consideration of the implications of these findings for current regulatory policy, as well as for alternatives to that policy.

First, however, it is necessary to inject a cautionary note on the uses of case studies in policy analysis. The case study approach is based on the belief that an intensive examination of the particular will yield insights about the general. Accordingly, this case study was not carried out to satisfy a unique curiosity about New Mexico, but to learn something of more general applicability about the enforcement of air quality regulations for stationary sources. To this end a case study can only be suggestive, not conclusive.

The usefulness of this case study, even for suggestions, depends on the extent to which the situation in New Mexico is typical of other states. There is, admittedly, some room for doubt on this point. It is reasonable to expect, for example, that the relative paucity of sources might make it easier for New Mexico authorities to tailor regulations to specific sources. This lack of specificity in other states might require that their authorities must use discretion in translating regulations into individual source requirements, thereby opening the door for negotiation. Even so, after

the source requirement has been established, the problems of maintaining

continuous compliance is the same in all states: Emissions must be moni-

tored and violations corrected. There is no reason to think that the

expense of source surveillance or its frequency will be notably different

in New Mexico.

Likewise, the supposed lack of a serious air quality problem in New

Mexico might be expected to contribute to a certain slackness in pursuing

regulatory objectives not found in other states. But this proposition is

dubious. For one thing, although ambient air quality in New Mexico is

generally better than national norms, it is incorrect to say that that

there is no cause for concern. The extraordinary visibility found in the

Southwest has deteriorated somewhat in the past few decades, a development

that has been linked to gradually increasing SO_2 emissions.[1] Besides, the

limited evidence available from other states suggests that EIA's affinity

for voluntary compliance and associated reluctance to seek legal remedies

for violations is far from unique. Although New Mexico is unusual in

having voluntary compliance required by statute, almost all states practice

it in fact.[2]

Even if New Mexico is perfectly representative, however, it is dangerous

to draw conclusions from only one observation. The "conclusions" of a case

----

[1]Trijonis, John and Kung Yuan, "Visibility in the Southwest: An Exploration of the Regional Data Base." U.S.E.P.A., Research Triangle Park, N.C., EPA 600/3-78-039.

[2]See "Profile of Nine State and Local Air Pollution Agencies," U.S. E.P.A., Office of Planning and Management, September 1980.

study are in reality hypotheses, which can then be tested in a more
general analysis. In this particular case, the data are sometimes so
sketchy that these conclusions must be regarded as arguable even with
respect to New Mexico.

Conclusions

The purpose of source enforcement, obviously, is to promote compliance with emission regulations. However, compliance is a much-abused
word, and it is essential to distinguish between initial and continuous
compliance. Initial compliance means that a plant has demonstrated
the ability to comply with the regulation. Even after achieving initial
compliance, however, a plant may not be in continuous compliance--in
the sense of meeting the regulation 100 percent of the time--because
of inadequate maintenance or improper operation of installed equipment
or variations in production variables. This distinction has a bearing
on every important aspect of the enforcement problem.

Surveillance. Determination of initial compliance is relatively
easy, inasmuch as it usually involves only a source test after the requi-
site equipment has been installed. Determination of continuous compliance
is infinitely more difficult. With few exceptions, the availability of
data on actual emissions from sources is such that one can have only a
rudimentary idea of the performance of sources, and therefore of their
continuous compliance with the regulations. Because continuous emission

monitoring technology is not yet well developed,[3] the only way of getting

direct information on source emissions is by visiting the plant. Source

surveillance is so expensive that plants are visited very infrequently--

in most industrial categories no more than once a year. Moreover, most

visits fail to get a reading of actual emissions because of the expense of

conducting a source test. Instead, the typical visit is an engineering

inspection of abatement equipment, together with visual observation of

stack and fugitive emissions. Obvious violations of the regulations can

be determined in this way. A third cause of difficulty in making infer-

ences from emission data is the fact the surveillance visits that are made

may not be made at "representative" times. Surveillance visits made in

response to complaints may find source performance which is worse than

usual. On the other hand, sources are generally notified beforehand about

routine visits, and this may bias the data in the opposite direction.

This rather bleak view of the adequacy of data on continuous com-

pliance must be tempered in several ways. First, continuous monitors

may in the future provide better information on continuous compliance.

Second, those sources and pollutants for which emissions can be deter-

mined by materials balance provide data on average emissions, at any rate.

In New Mexico average SO_2 emissions from copper smelters and natural gas

processing plants were determined in this manner. However, a materials

balance at best gives monthly averages of emissions, and is therefore not

---

[3]No source is required by New Mexico regulations to operate continuous
monitors. Federal New Source Performance Standards require continuous
monitors for some categories of sources, among them utility steam generators, but the data from these monitors cannot at present be used for
enforcement purposes.

much help in determining daily emissions, which is how the regulations are written. Third, surveillance is inexpensive for sources for which performance is determined by opacity reading, (such readings provide a numerical index of performance, even though the actual mass of emissions is not measured). Woodwaste burner emissions were determined in this way. Fourth, plants can be required to conduct source tests themselves on a routine basis. For large plants the cost of frequent source tests would be small relative to the costs of compliance, and therefore would not be particularly burdensome. At the Four Corners Power Plant, for example, relatively frequent tests were conducted on units 4 and 5.

Notwithstanding the scarcity of information, enough does exist to suggest that air emission regulations are frequently violated. Because many of the violations are determined by means of inspections and not source tests, it would be difficult if not impossible to estimate actual emissions in the aggregate. If this pattern of frequent violation is representative of the national situation, then national emissions may well be considerably underestimated.

Exemptions from regulation. Not all emissions in excess of regulation are considered "violations". In New Mexico a source may also obtain a variance which allows the source to exceed the emission standard, usually for a specified length of time. Variances are granted for any of several grounds, mostly concerned with the technical or, more commonly, economic feasibility of complying with the regulation. Somewhat like the variance is the assurance of discontinuance, in which a source found in violation of a regulation is allowed a grace period to return to compliance. An assurance of discontinuance is somewhat easier to get than a variance because no public hearing is required, but the duration is typically much shorter. Almost 40 percent of the stationary sources in New Mexico have sought variances or assurances of discontinuance at one time or another, but the variation among source categories is enormous. For example, while very few asphalt processing plants or gas-fired utility boilers sought them, almost every nonmetallic mineral plant did.

Excess emissions are also forgiven during startup, shutdown or malfunctions, providing such events are reported to the EIA. Allowing short-term exemptions of this sort is a very common practice among air quality agencies, presumably because excess emissions are often very costly to prevent. However, as regulations on "normal" emissions become more stringent, it may become cost-effective to regulate these "excess" emissions. In Chapter 4 it was argued that the point had been reached for an important class of sources, coal-fired electric power production. Using data on startups from Units 4 and 5 of the Four Corners Power Plant, it was found that the cost of eliminating startup emissions (by burning natural gas during startup) was less on a per-pound-removed basis than the marginal cost of EPA's recently revised New Source Performance Standards.

Incentives for Compliance. The enforcement philosophy of New Mexico's Air Quality Control Act is embodied in the "voluntary compliance" process. According to the statute, those found in violation of a particular regulation must be given a chance to return to compliance voluntarily before legal sanctions can be applied. When a violation is discovered, it is usually sufficient for the EIA to send a letter to the source pointing

out the fact, whereupon the source quickly returns to compliance. If
the source cannot or for some reason would prefer not to achieve compliance immediately, it can apply for a variance or an assurance of discontinuance, two formal mechanisms to permit a source to operate for awhile
with emissions in excess of standards without violation of the regulation.
Occasionally sources are requested to make some sacrifice (for example,
a curtailment of output) to hold down emissions during such periods.

By itself, the voluntary compliance procedure offers no incentives
for compliance; it must be supported by the threat of sanctions. Thus,
voluntary compliance is not so much an enforcement process as a
negotiating process. The negotiations involve what a source found in
violation of regulations can and cannot be expected to do with respect
to improving emission performance. This, of course, puts a premium on
technical information, which gives the source a couple of important bar-
gaining advantages. First, the agency must rely mainly on the source
for information about its production processes. In addition, negotiators
for the source will necessarily be more familiar with their own process
than will negotiators for the agency, who in all probability will have
to be familiar with many sources in a variety of industrial categories.
This disparity will very likely be exacerbated by the rapid turnover
and consequent low experience level of agency staff.[4]

Even though voluntary compliance is not, properly speaking, an
enforcement mechanism, it does condition the practice of enforcement in

[4]In New Mexico, this problem was mitigated by the fact that the head
of the enforcement section of the Air Quality Division has been with the
Agency since the first regulations written in 1970.

two important ways. First, because formal sanctions, i.e., court actions,
are involved only when the negotiations about voluntary compliance break
down, they are reserved for those deemed guilty of bad faith or outright
defiance of the regulations. The decision to threaten a source with
sanctions comes only after the EIA has given the source every opportunity
to achieve voluntary compliance. The actual use of formal sanctions has
been quite rare. In fact, only five cases have been referred to the
State Attorney General for prosecution since the first regulations were
promulgated in 1970.

The fact that sanctions are invoked to punish uncooperativeness means
that their use will bear little relationship to the magnitude of violations
of the regulations. In contrast, existing models of the enforcement process usually assume explicitly that the size of the fine will depend on
the magnitude of the violation,[5] which is the product of the amount by
which the emission standard is exceeded and the time out of compliance.
While lack of cooperation may be correlated with the duration of a violation, it bears no discernible relationship to rate of emission discharge.

Lack of cooperation may be a necessary condition for the imposition of sanctions, but it is not sufficient. All the industrial sources
which have been taken to court for violation of air pollution regulations
in New Mexico have been woodwaste burners. While this fact may perhaps
be accounted for by some unique attitudes among sawmill operators, this
explanation is not very plausible. The behavior of such operators has

---

[5]For example, Paul B. Downing and William D. Watson, "Enforcement
Economics in Air Pollution Control," Washington, Environmental Protection
Agency, EPA-600/5-73-004, December 1973.

not been appreciably more obstinate or uncooperative than some sources
in other categories. More likely the difference is due to the amount
of evidence available. Because woodwaste burners are regulated by
a percent opacity standard rather than a mass emission standard, determination of performance is much less expensive than for other types of
sources. Therefore, surveillance is much more frequent (about every
other month rather than once a year). With more frequent surveillance,
it is easier to construct a record of frequent violations, which would
support a claim that the source in question has failed to come into
compliance voluntarily. The conclusion is that a lack of evidence makes it
difficult to apply sanctions.

A second noteworthy feature of voluntary compliance is that it provides a source with no incentive to comply with an emission regulation
until after a violation has been discovered. Because a source must by
statute be given a chance to come back into compliance voluntarily, it
cannot be penalized for a first violation. Moreoever, the fact that
sanctions are employed only as a last resort means even repeated violations
would occur before formal sanctions are applied. This hardly provides
much incentive for careful maintenance of abatement equipment.

In fact, the rarity with which court action has been pursued raises
skepticism over the value of formal sanctions as incentives. Nonetheless,
there must be some incentives for compliance with air quality regulations.
As noted in Chapter 5, the EIA has been largely successful in forcing
initial compliance with the regulations, even though it took years to do
so for some plants. As for continuous compliance, the fact that many
plants are frequently out of compliance should not be allowed to obscure

the fact that they are even more frequently _in_ compliance. Even after

initial compliance has been achieved, continuous compliance can be costly,

and without some incentive, who would take the trouble? Identifying the

real incentives for continuous compliance is not easy. Perhaps one is

the threat of legal action, however remote. Perhaps uncertainty over the

reaction of the legislature and the electorate to a massive disregard

of the regulations is an incentive. Perhaps some plants comply simply

because they are required to do so by law.

Whatever these incentives are, sources have responded to the air

quality regulations in ways that are not surprising from an economic point

of view. That is, a plant faced with regulation has tended to choose

a course of action that minimizes its cost. Occasionally, this means that

the plant will come into compliance with the regulation before it becomes

effective. But if the New Mexico experience is any indication, plants will

usually seek some measure of regulatory relief, either by

> (i) getting a variance or assurance of discontinuance;
> (ii) challenging the regulation in court; or
> (iii) attempting to amend the regulation.

To explain whether a plant will seek relief and, further, to explain which

of the above alternatives will be pursued by a plant requires a considera-

tion of the costs of the above actions and their benefits.

The benefits of regulatory relief to a plant consist of the avoided

cost of installing and operating abatement equipment. Thus, the benefit

of a variance or assurance of discontinuance is a delayed investment in

equipment. These benefits are reasonably certain, inasmuch as so few

applications for variances and assurances of discontinuance were denied

in New Mexico. (Because a variance is usually of greater duration, it

is probably of more value.) The benefits of a successfully challenged
or amended regulation include not only delay, but possibly a less stringent regulation as well, although this is by no means certain. Because
an amended regulation will apply to all sources subject to that regulation, the source seeking the change will be unable to capture all the
benefits.

On the other hand, the costs of seeking regulatory relief are primarily administrative, consisting mainly of the resources devoted to correspondence and negotiation with the agency, including legal fees. In some
cases the source might suffer from adverse publicity, and there is also
the possibility that the source might come to be considered uncooperative,
and therefore not complying voluntarily.[6] These costs are extremely difficult to quantify, and it is doubtful whether a plant itself would have
more than a rudimentary notion of the cost of seeking regulatory relief.
Nonetheless, it appears that two qualitative assertions are possible.
First, responses above are probably listed in order of increasing cost.
It is obvious that obtaining a variance or assurance of discontinuance
requires less effort than challenging or amending the regulation. The
assurance is probably easier to execute than the variance because the
latter requires a hearing. The choice between a court challenge and an
amendment will probably not be made on the basis of cost, but rather on
the source's perceptions of the chances for success. The second point
to make about these costs is that there are likely to be considerable

------

[6]For Kennecott, the other large source which was successful in having
its regulations amended, economic considerations may have played a role.
Because of the depression in the copper industry the threat of shutdown,
and consequent loss of employment, was credible.

economies of scale. While there may be some relationship between the size of a plant and the amount of negotiation required to secure regulatory relief of any kind, it is almost certainly less than a linear relationship. Moreover, these economies of scale are likely to be more pronounced than the economies of scale in complying with the regulations.

These benefit and cost considerations, together with the empirical evidence from New Mexico, suggest several hypotheses about the response of sources to air quality regulations. First, the higher a plant's abatement costs, the more likely the source is to seek regulatory relief, and furthermore the more likely it is to attempt to change the regulation rather than simply delay its implementation. Second, the "public good" aspects of amended regulations within the community of plants suggest that court challenges or amendment attempts will be sought by either trade groups or sources subject to regulations applicable to only a few sources. Third, the larger the source, the more likely it is to seek regulatory relief. This follows because as sources become larger the cost of abatement increases faster than the administrative cost of postponing or challenging the regulation. The limited evidence from New Mexico seems to favor this hypothesis over the more common explanation of the apparently greater success of large sources in seeking more favorable regulations, namely, greater political or economic clout. In particular, the Four Corners Power Plant was very successful in postponing and amending regulations but, because virtually all the power produced is exported, possesses little discernible economic or political power.

The hypotheses above have sought mainly to explain behavior with respect to initial compliance, but a source's continuous compliance record

is also affected by economic considerations. In this instance the
benefits of noncompliance are that it permits costs of operating and
maintaining the equipment installed to meet the regulation to be reduced.
Therefore, if the operating costs are high relative to other costs, then
we can expect to see frequent violations of the regulations even after
initial compliance has been achieved. Thus the frequency of violations
found among both asphalt processors and woodwaste burners, both with
high abatement operating costs, was 39 and 33 percent, respectively.
The lack of better performance among woodwaste burners further suggests
that formal sanctions will not be effective as incentives for continuous
compliance. Due to the cumbersomeness of legal processes this may be
true in general, but more likely the effectiveness of sanctions is
limited by voluntary compliance, which forbids the application of
sanctions until after a source in violation has been given a chance
to come back into compliance.

The response of sources does not depend entirely on the costs of
compliance; the cost of noncompliance may also play a role. In
particular, we have that _ceteris paribus_, sources will have better
compliance records in urban areas, where violations are more likely
to be observed and more likely to generate adverse publicity. Thus a
number of sources in Grants, Albuquerque, Alamogordo, and other cities
have exemplary compliance records (although many do not). For public
relations and to avoid nuisance actions, some even instituted significant control of emissions long before regulations were adopted
(obviously, this is a consideration only for visible and odorous
pollutants.)

Policy Implications

The experience in New Mexico supports the view that plants and firms

subject to air pollution abatement regulations have strong incentives to

attempt to postpone their compliance with those regulations. This problem

was widely noted at the time the Clean Air Act was amended in 1977 (P.L.

95-95), and in response a "noncompliance penalty" provision was enacted

(Sec. 120).[7] This section requires that a plant which delays compliance

is liable for an administrative penalty assessment (without resort to

formal legal procedures) equal to the economic benefits of delayed compli-

ance. In July, 1980 EPA promulgated regulations to implement this section.

So far there is no experience with the penalty, so it is unknown whether it

will achieve its legislative intent. In any event, it is unlikely that the

regulations will affect continuous compliance. Although Section 120 refers

to compliance generally, the practical problems of estimating the penalty

for a continuous violation are severe.

These findings have important implications for the allocation of

public resources to achieve ambient air quality objectives. The present

regulatory approach to air quality management consists of several activi-

ties, including air quality monitoring, writing source-specific regulations

---

[7]The model for Section 120 was the Connecticut Enforcement Act of
1973 (P.L. 73-665). Although few noncompliance penalties have been
assessed, it is possible that mere existence of an administrative penalty
can provide incentives for compliance. At any rate this is the belief of
State officials in Connecticut. I am not aware of any quantitative studies
comparing compliance in Connecticut with other states, however. See U.S.
E.P.A., Connecticut Enforcement Project Report, Washington (1977).

to achieve ambient objectives, and enforcement of those regulations. The very limited data on the actual performance of sources, as well as the high frequency of violation found in those data that do exist, suggest that more resources could advantageously be shifted from the current practice of writing ever more stringent regulations for new sources to the more vigorous enforcement of existing regulations. Indeed, as regulations become more stringent, the benefits of noncompliance increase, making enforcement all the more problematical.

It also may be important to consider how resources should be allocated among the various enforcement activities. At present in New Mexico, almost all enforcement resources are devoted to surveillance and negotiation, rather than providing compliance incentives. Although this may suggest that perhaps greater attention should be paid to incentives, it is doubtful whether the existing quality and quantity of performance data will support more vigorous enforcement of the regulations. As noted above, the application of formal sanctions requires a lot of data on the performance of sources. If the existing surveillance effort does not generate enough data to support the imposition of sanctions--and at an average of one surveillance visit per source per year there is plenty of reason to believe it does not--then this question of resource allocation in enforcement is premature. It is possible that the existing level of surveillance makes the voluntary compliance approach inevitable.

These results also have implications for the allocation of enforcement resources among sources. Clearly, those sources for which abatement operating costs have the greatest impact are the ones which benefit most from noncompliance, and should therefore be under the closest surveillance.

129

In addition, different sources may have very different impacts on air quality. In New Mexico, and probably almost everywhere else, the majority of emissions come from a small number of sources. For example, in New Mexico more than 80 percent of SO_2 emissions came from the Kennecott Smelter and the Four Corners Power Plant. This preponderance should probably be reflected in source surveillance patterns.

The difficulties uncovered here in the implementation of existing air quality policy may have parallels for other, as yet untried policies. For example, the problems which arise when implementing an effluent charge policy (involving the determination and collection of the annual or monthly payment for each source) are analogous to the problem of enforcement under a policy of effluent standards. The tasks of enforcement are first to determine what the emissions are and then to take action based on that information, and these tasks remain under a policy of charges. As this report makes clear, these tasks are significant sources of uncertainty, in a regulatory regime. Effluent charges remove one of these sources of uncertainty, because once the emission information is known the calculation of the firm's payment is mechanical. Under perfect information the charge approach exacts a payment for each unit of emission discharged; therefore, the problem of properly weighing equipment performance and reliability, which is inherent in the effluent standards approach, is avoided. In our view this is a significant virtue of effluent charges which is often overlooked.

Nonetheless, the problems of determining emissions remain, and may, in fact, be more troublesome under an effluent charge policy than under direct regulation because more information may be required. Under an effluent charge policy, if the payment is to be based on a source's actual emissions, then there must be some way for those emissions to be estimated. (With effluent standards, one only need to know whether the standard is exceeded, which can often be done without estimating emissions.) If feasible, the techniques of continuous monitoring or materials balance will provide the required estimates.

Continuous monitoring and materials balances will cover only a minority of situations, respectively omitting small facilities for which the cost of continuous monitoring would be onerous, and those pollutants, such as NO_x, for which materials balances are infeasible. To estimate actual emissions in those situations, one would need at a minimum a set of randomly drawn effluent samples from which an average could be calculated. For airborne emissions, this means a set of source tests will be required, each costing $1,000 to $5,000.[8] While a trivial expense for the large plant, this could be a serious burden for a small operator. If the rate of pollutant discharge is both highly variable and unpredictable, then the calculation of the effluent charge from monitoring data takes on aspects of a lottery. For example, consider a source which in response to an effluent charge reduces its pollution discharge by 95 percent, and suppose that the reliability of equipment installed for this purpose is 95 percent. The other five percent of the time the discharge is at the

---

[8]This is only the cost of the test. It does not include any costs the source may incur in idled personnel or reduced output while the test is being prepared or conducted. The cost of surveillance would be the same regardless of whether the surveillance were done by the air quality agency or through "self-reporting," as recommended by some commentators (e.g., Anderson, et. al., _Environmental Quality Through Economic Incentives_, Johns Hopkins Press for Resources for the Future, 1978). To be sure, the distribution of the cost would be quite different.

"uncontrolled" level, or what it was before the imposition of the charge.

Thus, the average reduction at the source is about 90 percent. Suppose

the payment is determined from instantaneous emissions calculated twelve

randomly selected times a year, which is quite frequent by the lights of

current regulatory policy. Under these assumptions the plant's bill would

very probably vary considerably from what it would have to pay if its

emissions were known with certainty, as shown in the following table:

Probability	Effluent charge bill as a percent of "correct" payment
0.54	51%
0.34	133%
0.10	213%
0.02	295%

Variation of this magnitude would probably be viewed as unacceptably

capricious.

For this reason it appears reasonable to expect that effluent charge

payments would be based not on actual surveillance but on expected or

nominal emissions, at least for sources or pollutants for which neither

continuous monitoring nor materials balance is feasible. One possible

solution is to base the effluent charge on a prescheduled source test

conducted annually, with occasional inspections during the year to

enforce compliance with the test result. This is not dissimilar to the

approach to enforcement found under the current regulatory policy, and

not surprisingly shares its defects with respect to incentives, creating

a bias in favor of equipment or techniques which achieve high instantan-

eous performance at the expense of reliability.

There is an additional perverse incentive which needs to be mentioned. A source would probably be better off if it were not one of those required to install continuous monitoring, but instead could get by with an annual source test as discussed above. The reason is that during a source test the plant can take pains to ensure that its emissions are as low as possible, and in that way show emissions which are atypically low. This would obviously provide a barrier to improvements in continuous monitoring technology. Moreover those sources which already were required to have continuous monitors would have an incentive to increase the monitors' unreliability or inaccuracy. To reverse this disincentive the agency would have to do something to make it more expensive for sources to be exempt from a continuous monitoring requirement than to be subject to it. It could do this by charging a premium to those in the "exempt" class. But this would be likely to penalize small sources disproportionately. Whatever the merits of such a penalty, we have grave doubts concerning the political feasibility of an effluent charge which is regressive with respect to plant size.

Thus, the difficult problems of surveillance and enforcement do not disappear when an effluent charge is substituted for a set of standards, unless some reliable means of continuous monitoring is available. But, of course, the availability of continuous monitoring makes the task of enforcing standards much easier as well.

For Product Safety Concerns and Information please contact our EU
representative GPSR@taylorandfrancis.com Taylor & Francis Verlag GmbH,
Kaufingerstraße 24, 80331 München, Germany

Batch number: 08158242

Printed by Printforce, the Netherlands